What in the World is DNA?

What in the World is DNA

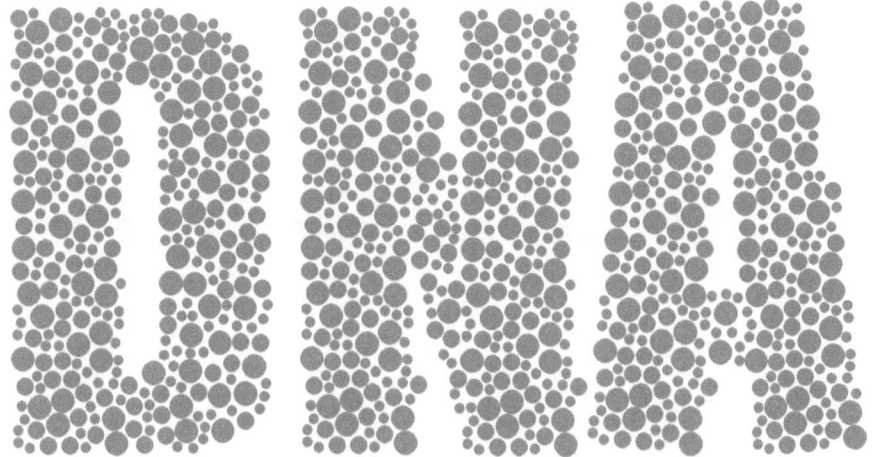

WRITTEN AND EDITED BY

Dr Austin Mardon, Annilea Purser, Benjamin Turner, Keshikaa Suthaaharan, Paawan Virdi, Priyanshu Mahey, Ghulam Aisha, Nawshin Haq, Neha Saroya, Nazihah Alam, Poojitha Pai, Hayley Zhong, Juwairia Razvi

Golden Meteorite Press
2021

Copyright © 2021 by Austin Mardon
All rights reserved. This book or any portion thereof may not be reproduced or used in any manner whatsoever without the express written permission of the publisher except for the use of brief quotations in a book review or scholarly journal.
First Printing: 2021

Typeset and Cover Design by Jenna Usselman

ISBN 978-1-77369-230-2
Golden Meteorite Press
103 11919 82 St NW
Edmonton, AB T5B 2W3
www.goldenmeteoritepress.com

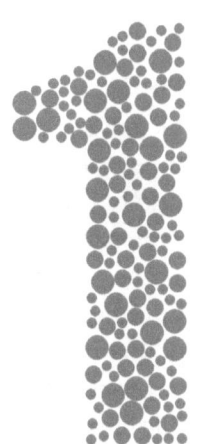

What Did People Think About Genes Before DNA was Discovered?
ANNILEA PURSER

"Genes are like the story, and DNA is the language that the story is written in."
-Sam Kean

Commonly, when one reflects on historical scientific discoveries in the 20th century, they consider findings related to quantum mechanics or even nuclar fission. However, after the second world war, genes and their function in genetic inheritance became a prominent topic for scientific discourse. Although deoxyribonucleic acid (DNA) was first identified by Friedrich Miescher in the late 1860s, the popularization of the field of genetics was sparked by discoveries about the structure and function of DNA by researchers James D. Watson and Francis H.C. Crick in 1953 when they clarified the role of DNA in housing genes (Nature Education, n.d). With the discovery of DNA's function and its popularization in the field of science taking place relatively recently, prior to delving into the specifics of the discovery itself, it is imperative for one to understand how scientific researchers, scholars, and the general public viewed genes before the era of Watson and Crick. Thus, this chapter will chronologically look at the evolution of genetics prior to the popularization of DNA through 7 key events; the invention of the single-lens optical microscope, the description of the cell, Charles Darwin's On The Origin of Species by Means of Natural Selection, Gregor Mendel's fundamental laws of inheritance, Heinrich Wilhelm Gottfried von Waldeyer-Hartz's chromosome description, Theodor Boveri's findings, and Thomas Hunt Morgan's locating of genes.

The Single-Lens Optical Microscope
The creation of the single-lens optical microscope was paramount in scientists

understanding cells and the field of scientific research as a whole, thus meaning that it was critical to the development of perceptions about genetics. Although it cannot be conclusively traced back to him, it is widely accepted that the single-lens optical microscope was invented by Zacharias Janssen with the help of his father, the renowned Hans Janssen, in the 1590s (Hedlund, et. al, 2015, p.1). The Janssens developed the microscope by placing multiple lenses in a tube, which had never been done before (Hedlund, et. al, 2015, p.5). As one can imagine, the first creation of the single-lens optical microscope was not perfect in any sense of the term, especially considering what is accepted as effective scientific instruments in the present day. However, it was a major scientific breakthrough in beginning to be able to visualize intracellular structures (Hedlund, et. al, 2015, p.1). In extension of this, although no direct observations made from the single-lens optical microscope were published, future researchers would utilize their creation and embrace the microscope structure by the Janssens as a legitimate scientific instrument which would later be used to make observations about genes.

Cell Ideation & Theory

One individual who utilized the single-lens microscope to make important contributions to the study of genes was the English multidisciplinary scientist Robert Hooke. Robert Hooke, a scientist with contributions spanning over many focus areas including mathematics, optics, mechanics, architecture, and astronomy, was a naturally curious man (O'Connor et. al, 2002). This curiosity led Hooke to wonder about the structure of corks as he o served their natural buoyancy and spring-like nature (O'Connor et. al, 2002). With this, Hooke performed his well-known cork experiment where he sliced a slim piece of cork and placed it under the single-lens optical microscope where he then observed that the cork looked like a piece of honeycomb (Fentress, 2019). Within this honeycomb pattern, Hooke was able to conclude that in every cubic inch of cork was approximately twelve-hundred million small sections separated by thin walls (Fentress, 2019). He referred to these small sections as cells, thus coining the term (Fentress, 2019). His curiosity was therefore appeased as he could safely speculate that corks float because of air within each individual cell (O'Connor et. al, 2002). This discovery by Hooke was critical to scientists being closer to understanding the structure of human life, and in-extension genetics as "knowing the structure of cells and the processes they carry out is necessary to understanding life itself" (Grewal et. al, 2021). Soon after Robert Hooke's discoveries, the Dutch scientist Anton van Leeuwenhoek began working on strengthening the single-lens optical microscope so that he could better observe human cells and bacteria (Grewal et. al, 2021). Similar to Robert Hooke, Anton van Leeuwenhoek was curious-by-nature which led him to take up a hobby of grinding lenses (Britannica, n.d). Although his methods are described as being highly flawed by contemporary standards, Anton van Leeuwenhoek was the first individual to study human cells under a microscope which motivated others to begin doing the same (Britannica, n.d). In fact, by the early 1800s many scientists were inclined to study diverse types of cells in a diverse number of ways. This prompted two German scientists, Theodor Schwann and Matthias Jakob Schleiden to perform their own research on cells and proposed that cells are in fact the basic ingredient for all living things (Grewal et. al, 2021). Rudolf Virchow agreed with the findings of Schwann and Schleiden, and furthered them in his groun breaking observations. Virchow utilized an even stronger type of microscope than his predecessors (namely, the aforementioned Janssens and Robert Hooke) to discover that cells are active by nature and divide to form new cells (Grewal et. al, 2021). He then even found that all living cells are only generated by other living cells (Grewal et. al, 2021). The combination of ideas from Schwann, Schleiden, and Virchow resulted in the infamous cell theory that rests on three important, yet distinct concepts (Grewal et. al, 2021). The cell theory proposes that (1) all organisms are made of one or more cells, (2) that all life functions of living things occur within cells, and (3) that all cells come from already existing cells. At this point in history, the idea of genes was not yet being explored and genetics as a field was not yet conceived, however, the ideation of cells and cell theory were critical steps in advancing scientific research to the discovery of genes.

Darwin's Genetic Inheritance Introduction

By 1859, scientists had a general understanding of the function and idea of cells, but it was Charles Darwin who applied this knowledge to the concept of trait inheritance (Stanford Encyclopedia of Philosophy, 2019). In fact, the conclusions that Darwin drew are considered to have had massive impacts within scientific research, which are arguably unmatched by any other historical scientific figures (Stanford Encyclopedia of Philosophy, 2019). Darwin's work, On The Origin of Species by Means of Natural Selection, presented one of his most influential theories: natural selection (Stanford Encyclopedia of Philosophy, 2019). Within this theory, Darwin argued that evolution is dependent on the presence of heritable aspects of any given species. Although other theories were present at the time, Darwin's was most impactful due to the fact that it emphasized factors controlling populations and did not describe evolution as an adaptive process (Charlesworth et. al, 2009, p.757). The work of Darwin acted as a precursor to the field of genetics, due to the fact that it presented the idea of cells being responsible for trait inheritance. In particular, in On the Origin of Species by Means of Natural Selection, Darwin presented the idea of the transmission of traits down several generations of pedigrees (Charlesworth et. al, 2009, p.758).

However, Darwin's work did not go uncontested. In fact, his theory on natural selection was largely dismissed or refuted by theorists who held a wide array of criticisms (Charlesworth et. al, 2009, p.757). These criticisms were most frequently rooted in Victorian religious beliefs and refusal to believe that man evolved from monkeys (History Extra, n.d). Unfortunately, at the time, Dawin failed to provide solid empirical evidence circumstanting that his theory was "right". He did, however, attempt to refute the arguments of other theorists. For example, in the aforementioned example of the transmission of traits down several generations of pedigrees, Darwin utilized assistance from the mathematical physicist Sir George Stokes to show that the probability of transmission of traits was higher than not (Charlesworth et. al, 2009, p.757). Regardless of the contestation of his ideas, it is a widely accepted idea that Darwin brought forth the idea of trait inheritance which would act as a stand-in for the field of genetics and begin shaping public opinion on genes.

Gregor Mendel's Laws of Heredity
Although Darwin's claims to evolution and trait inheritance were largely dismissed due to a lack of circumstantial evidence, another researcher came to the rescue just a few years later: Gregor Mendel (original name, Johann Gregor Mendel). Similar to his predecessors like Robert Hooke, Gregor Mendel was a multidisciplinary academic, with interests in a wide range of subject matters such as botany (Olby, n.d). In his early life, Mendel was a student at the Philosophical Institute of the University of Olmutz (Olby, n.d). This is where Mendel began his interest in trait inheritance; under the cell theory enthusiast Franz Unger, Mendel began studying the physiology of plants and became well-acquainted with examining cells under a microscope (Olby, n.d). Eventually, however, Mendel and his mentors felt it was necessary for him to continue receiving more formal education, thus resulting in his move to Vienna (Olby, n.d).

In approximately 1854, Mendel took his interest in studying cells to a higher level and proposed an experimental program on hybridization to his mentor, abbot Cyril Napp (Olby, n.d). The aim of this experimental program would be to better understand the transmission of hereditary traits in successive generations of hybrid progeny (Olby, n.d). In other words, Mendel wished to examine how certain traits (i.e height) are passed on (Olby, n.d). To Mendel's benefit, abbot Cyril Napp agreed to supervise this research (Olby, n.d). Prior to Mendel's inquiries, scientists performed little research on hybrid progeny, but did observe that progeny of fertile hybrids typically return to the origin species (Olby, n.d). This resulted in the previously-accepted conclusion that hybridization would not be a system for multiplying species, however, that certain rare cases may result in fertile hybrids reverting which are to be called constant hybrids (Olby, n.d)). These pre-Mendel findings proved to be important on a number of grounds. For one, it confirmed the beliefs of plant owners and animal breeders who had previously shown that crossbreeding can result in a diversity of new forms, not just one (Olby, n.d). These findings were also viewed as important as certain landowners were concerned. For example, the abbot ofthe monastery was concerned about the wool of its Merino sheep as they were competing with Australian wool qualities (Olby, n.d).

Mendel's research proposed to conduct a two-year long examination of edible peas (Pisum Sativum) (Olby, n.d). He chose edible peas as his test matter due to their distinct varieties, "ease of culture", ease to control pollination, and due to the high volume of seed germinations already known (Olby, n.d). Mendel tested 34 varieties of the edible peas for constancy of 7 potential traits. For example, one trait could be size (big/small). He referred to these trait opposites (i.e big and small) as contrasted characters which could either be dominant or recessive (Olby, n.d). First, utilizing the numerical correspondence $F1$, or first generation of hybrids, Mendel simply observed dominant versus recessive traits (Olby, n.d). Then, using the numerical correspondence $F2$, or second generation, he once again observed dominant and recessive traits, and noted that the recessive traits reappeared in a 3:1 ratio (Olby, n.d). Further, using $F3$, or descendants, Mendel found that only ⅓ of the variants were considered a "true" breeding whereas ⅔ were of hybrid constitution (Olby, n.d). This 3:1 ration was then rewritten as 1:2:1 meaning that, overall, 50% of the second generation were true breeding and 50% were still hybrid (Olby, n.d). This finding was, and is still considered Mendel's most famous finding as it championed the field of genetics (Olby, n.d). With a newfound confidence in his research, Mendel then tested another gene-related hypothesis: that every trait is transmitted independently. For example, Mendel wished to discover whether the traits of size (big/small) and colour (green/yellow) were connected, or if they are separate traits altogether. Mendel proved that each trait transfers separately by crossing a number of the peas (Olby, n.d). This finding was later coined as the process of independent assortment in genetics.

The research performed by Mendel was also summarized by later researchers in a 3-part theory titled 'Mendel's Laws of Heredity' which simplify the aforementioned points (DNA Learning Centre, n.d). The theory states (1) The Law of Segregation where every trait is defined by a gene paid which are randomly selected, (2) The Law of Independent Assortment where genes from different traits are separate, and (3) The Law of Dominance where organisms reflect gene forms that are dominant, not successive (DNA Learning Centre, n.d. Although Mendel's research is now considered to be so groundbreaking that he is referred to as the "father of modern genetics", when he first presented his research he was overwhelmingly overlooked as some individuals overlooked the evolutionary implication of his research (DNA Learning Centre, n.d). Regardless, Mendel's research furthered that of Darwin and truly championed the field of genetics as a whole.

Post-Mendel Development & Locating of Genes

Similar to how Mandel provided evidence to support the claims of Darwin, other scientists needed to provide further evidence and explanation of Mandel's work for it to be fully appreciated. Following Mandel, various scientists became comfortable with the cell theory and the idea that hereditary information is located in the nucleus (O'Connor et. al, 2008). However, the actual physical nature of this hereditary material was still unknown apart from the understanding that it is somewhere in cells. This is when, despite still having relatively poor-functioning microscopes, scientists like Walther Flemming began exploring the "fibrous network" in the nucleus which was termed the chromatic of "stainable material" (O'Connor et. al, 2008). Later, this term was coined by Heinrich Waldeyer as being chromosomes. Walther Flemming further observed that during cell division, the chromosomes shaped into thread-like bodies along their length during the mitosis stage (O'Connor et. al, 2008).

Theodor Boveri

Then, the German embryologist Theodor Boveri took Walther Flemming's findings and applied them to his own research. Utilizing cells from the roundworm species Ascaris Megalocephala, which are larger and more clear than the average cell, Boveri was able to provide the first ever evidence that chromosomes of germ cells (biological cells which are sexually reproducing organisms) provide continuity between generations (O'Connor et. al, 2008). This was a major scientific breakthrough as it provided proof and explanation that genetic inheritance happens through reproduction.

Walter Sutton

Following the research of Theodor Boveri was the American graduate student, Walter Sutton. Sutton, who understandably utilized a more advanced cytological model than his predecessors, discovered that it is possible to determine individual chromosomes in cells, particularly when they are in meiosis (O'Connor et. al, 2008). Sutton found this by studying the testes of the lubber grasshopper (Brachystola Magna) (O'Connor et. al, 2008). Further, he determined that it is possible to describe the configurations of individual chromosomes and that there are 11 pairs that can be distinguished just based on their dimensions (O'Connor et. al, 2008). Sutton also proposed that chromosomes have individuality which is maintained, even through generations. Overall, Sutton's work articulates the chromosomal theory of inheritance which was a critical step in the field of genetics and developing scientific perspectives on genes.

Thomas Hunt Morgan

The final individual to be discussed in this chapter is the American scientist, Thomas Hunt Morgan. Alongside his colleagues at Columbia University, Thomas Hunt Morgan located hundreds of fruit fly (Drosophila) genes and made multiple breakthrough discoveries about their genetic transmission (O'Connor et. al, 2008). In particular, he observed that one of his flies had rare white eyes which he discovered can be due to mutations (O'Connor et. al, 2008). Through testing various flies following this, he eventually learned that genes are located on chromosomes (O'Connor et. al, 2008). Thomas Hunt Morgan's largest contribution to the field of genetics is his confirmation of Mendelian laws of inheritance and his hypothesis that genes are located on chromosomes (O'Connor et. al, 2008).

Summary

From the invention of the single-lens microscope by the Janssens, all the way to Thomas Hunt Morgan's locating of genes, the evolution of the field of genetics has spanned over hundreds of years. What once was an unimaginable concept is now something that is considered a norm, and, despite it being common to overlook genetics as a major discovery of the twentieth century, it has changed the world of science in incomparable amounts. Circling back to the quote which was presented at the beginning of this chapter, "Genes are like the story, and DNA is the language that the story is written in."(Sam Kean), one can understand how the development of the field of genetics is imperative to understanding the story of human life. The following chapters of this novel will explore how the field of genetics discovered DNA, or the language of human life, and the historical, societal, and cultural impacts of it.

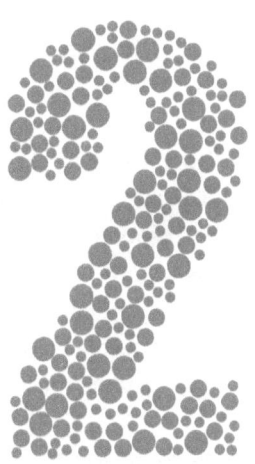

Who Discovered DNA? What is DNA?
Benjamin Turner

As with any scientific discovery, it is important to recognise that the process is messy; DNA is no exception. Many theories have been proven to be incorrect, and human understanding of the mechanisms involved in DNA and genetics has been redirected many times over. At times the research is messy in a more literal sense, with Friedrich Miescher beginning his work by scrubbing lymphocytes from soiled bandages to obtain the source material for his experiments (Dahm, 2008).

We should avoid being confused by the fact that today we use the terms genetics and DNA interchangeably as this was not always the case. Scientific debate about where genes were stored and how they were transferred between generations was not settled until Alfred Hershey and Martha Chase published their work about the reproduction of bacteriophages in 1952, and James Watson and Francis Crick discovered the double helix structure of DNA in 1953 (O'Connor, 2008). Until that point, many scientists believed that while DNA had a role in passing on genetic information, it was not complicated enough to be the primary mechanism at work. Many favoured the idea that proteins must be the carriers of genetic data because of the diversity and complexity of those molecules.

This chapter will discuss the researchers whose work made the largest leaps in our understanding of DNA and will provide a general explanation of what each one said about its structure and function. It will tell the story from when Miescher scrubbed bandages for lymphocytes to when X-ray images and physical models of the double helix were discovered. The aim is ultimately to consider how our understanding of DNA has evolved over time and answer the question: who discovered DNA and what exactly is it?

Who Discovered DNA?

The discovery of DNA is a complicated web of researchers all building on each other's discoveries with some of the largest leaps coming from Friedrich Miescher, Phoebus Levene, Erwin Chargaff, and James Watson and Francis Crick. This is by no means an exhaustive list of every scientist who contributed to the discovery, but in terms of developing our understanding of its existence and structure, these are perhaps the most notable.

The first person to isolate DNA was the Swiss researcher, Friedrich Miescher, in 1869 (Pray, 2008). History demonstrates that scientific discovery is not linear, and researchers may find themselves limited by their time and the resources available to them. Miescher is an excellent example of a scientist ahead of his time with his research, hampered and even sabotaged by the methods available to him.

In his early experiments, Miescher attempted to determine the chemical composition of cells, setbacks plagued his work He intended to discover how cells worked by analyzing their proteins, so he aimed to observe their properties and then classify them. Unfortunately the diversity of proteins was too great and the equipment and methods of the time were too elementary to complete this task. In the course of these early experiments, he developed a method for obtaining a crude precipitate of DNA; he was not sure what he had discovered but he was sure that this substance, extracted from within the nucleus of the cell, was not a protein and had not been previously identified. Miescher continued his experiments on this new substance and dubbed it "nuclein" (Aldridge, 2003).

By late 1869, Miescher had finished his initial analysis and had published his research by 1871. Miescher had discovered the building block of all life, but this fact would not become apparent for many decades. He continued his study of nuclein for most of his life while simultaneously participating in the scientific debate about how hereditary traits are passed on. At one point, Miescher wrote, "If one wants to hypothesize that a single substance specifically is the cause of fertilization in any way, then – without a doubt – one would have to think primarily about nuclein." (Dahm, 2008, p. 325). This idea that nuclein was the mechanism for fertilization did not persist for Miescher, however. Ultimately, he was unconvinced that any one substance could account for the range of diversity seen between or even within species and fell into the camp that argued proteins must be responsible for transmitting hereditary information. Despite being incorrect on this point, it is worth noting that he developed a hypothesis that came remarkably close to how information is stored chemically; he speculated that the arrangement of atoms within molecules could encode information at. He argued that these molecules could be made up of different specific geometric arrangements of their constituent atoms, most likely in the form of proteins. He further proposed that errors in individual molecules might be prevented from manifesting themselves in the developing embryo by the fusion of information from two germ cells during fertilization; this theory of course coming eerily close to how intact alleles from one parent can compensate for defects in the allele inherited from the other (Dahm, 2008).

In 1889, another researcher, Richard Altmann, renamed nuclein as nucleic acid (Dahm, 2008). This development frustrated Miescher because he had been quite clear on the acidic properties of nuclein and felt like other researchers were co-opting his work. Unfortunately for Miescher, his contributions to the field were overshadowed by other scientists because of his introverted nature. He was a poor promoter of his work and a perfectionist to boot, which lead to delays in publishing his work because he insisted on repeating many of his experiments to improve his methodology and secure better results (Dahm, 2008). As a result, Miescher's association with DNA is often overlooked today, and he felt that other researchers were overshadowing his work. It took nearly 75 years after his discovery for the importance of DNA to become clear to the world.

The next great leap in understanding DNA came from Russian physicist turned chemist Phoebus Levene (Aldridge, 2003). Working at the Rockefeller Institute in New York in 1919, Levene was the first to argue there are three major components of nucleotides (Pray, 2008). From his experiments with nucleic acids from yeast, he proposed first that nucleic acids were composed of a series of nucleotides and that a nucleotide was made up of one of four nitrogen-containing bases, a sugar molecule, and a phosphate group; this model was defined as polynucleotide. In this theory, the four possible nitrogenous bases are the pyrimidines: cytosine (C) and thymine (T), and the purines: adenine (A) and guanine (G) (Pray, 2008).

He also proposed what he called a "tetranucleotide" structure, wherein nucleotides were always ordered in the same or a very similar pattern. Unfortunately this proved incorrect, with scientists eventually realizing that Levene's tetranucleotide structure was overly simplistic and that the order of nucleotides along a strand of DNA was far more diverse (Pray, 2008).

Erwin Chargaff of Austria, was inspired by the work of Oswald Avery, who made a compelling case that DNA was genetic material; Avery's work is discussed in more detail later in this chapter. Chargaff expressed that what Avery discovered pointed the way to a host of subsequent discoveries (Pray, 2008). First, in 1949, Chargaff set out to see if there were differences in DNA among different species (Aldridge, 2003). In doing so, he disproved Levene's theory of a tetranucleotide structure, finding that each species differed in the amount of A, C, G, and T but that within a species, the proportions were identical. In other words, he discovered that nucleotides did not have a simple repeating pattern, and different species had different DNA compositions.

His next discovery is what we now refer to as base pairing rules. Chargaff recognized that the proportion of the pyrimidine T was always identical to the purine A, and the proportion of the pyrimidine C was always the same as the purine G. Simply put, T always binds to A, and C always binds to G. Despite the overall pattern being complicated and appearing random, the nitrogenous bases always paired in the same way. This rule became known as Chargaff's ratios (Aldridge, 2003).

The double helix structure of DNA was discovered by James Watson and Francis Crick in 1953 (Aldridge, 2003), but their work would not have been possible without the contributions made by various other researchers in the late 1940s, most notably Maurice Wilkins at King's College, London (Aldridge, 2003). Wilkins was curious about the long strands that DNA forms when pulled out of watery solutions with a glass rod. He wondered if this implied a regularity to the structure of DNA.

To test his theory, Wilkins began using X-rays to take pictures of DNA, but the images were difficult to interpret. In 1951 Rosalind Franklin joined Wilkins, bringing her expertise with X-ray crystallography (Aldridge, 2003). Franklin built a dedicated X-ray lab that produced the most accurate images of DNA at the time. Based on those images, she began to wonder if the molecule had a coiled, helical shape.

The advances made by Franklin and Wilkins paved the way for Watson and Crick to propose that DNA has a double helix structure. Their method for working this out seems quaint, but it allowed them to visualise the DNA molecule more intuitively than pencil and paper would. Specifically, they constructed the individual elements of nucleotides out of cardboard, then later metal and wood, and proceeded to work out how those pieces fit together physically on their desks (Pray, 2008). However, some early setbacks occured because their initial understanding of how thymine and guanine were configured was incorrect. They were also unsure whether the phosphate on their nucleotides went on the inside or the outside (Aldridge, 2003). After discussion with a colleague, American chemist Jerry Donohue, Watson made new cardboard cut-outs of the two bases with the phosphates on the outside. He wanted to see if a new atomic configuration would improve the model. Not only did the pieces fit together perfectly under the new arrangement with complementary base pairs matched up, and each pair held together with a hydrogen bond, but the model also reflected Chargaff's rules. Watson and Crick noted later that seeing the model with its new configuration immediately implied function; they predicted that DNA replicates itself during cell division. with the phosphate on the outside, there was an easy potential mechanism for copying genetic material.

There have been minor adjustments to Watson and Crick's model by subsequent researchers, but their primary findings remain intact. For example, Pray (2008) wrote the four major points they came up with were:

- "DNA is a double-stranded helix, with the two strands connected by hydrogen bonds. A bases are always paired with Ts, and Cs are always paired with Gs, which is consistent with and accounts for Chargaff's rule.

- Most DNA double helices are right-handed; that is, if you were to hold your right hand out, with your thumb pointed up and your fingers curled around your thumb, your thumb would represent the axis of the helix and your fingers would represent the sugar-phosphate backbone. Only one type of DNA, called Z-DNA, is left-handed.

- The DNA double helix is anti-parallel, which means that the 5' end of one strand is paired with the 3' end of its complementary strand (and vice versa). Nucleotides are linked to each other by their phosphate groups, which bind the 3' end of one sugar to the 5' end of the next sugar.

- Not only are the DNA base pairs connected via hydrogen bonding, but the outer edges of the nitrogen-containing bases are exposed and available for potential hydrogen bonding as well. These hydrogen bonds provide easy access to the DNA for other molecules, including the proteins that play vital roles in the replication and expression of DNA." (para. 12)

In 1953, Watson and Crick published their findings in an issue of the Nature journal. Franklin and Wilkins published their respective papers back-to-back with Watson and Crick in the same journal, and Watson, Crick and Wilkins won the Nobel prize for their work in 1962. Sadly, Franklin died of cancer in 1958 at the age of 37 (Aldridge, 2003).

One of the developments in the structure of DNA has been identifying three different styles of the double helix. The most common of these in living cells is the version proposed by Watson and Crick and is known as B-DNA (Pray, 2008). There is also A-DNA which is shorter and wider and has been found in dehydrated samples of DNA; A-DNA is rare in normal physiological conditions. The third is Z-DNA, which possesses a left-handed spiral instead of the right-handed spiral typically seen. Z-DNA is transient, only existing in response to specific types of biological activity. Discovered in 1979, Z-DNA was largely ignored. Importantly, however, recent data shows some proteins form strong bonds with Z-DNA, pointing to a possible biological role in protection against viral disease (Rich & Zhang, 2003).

Discovering that DNA is Genetic Material

As mentioned earlier, it was not always known that DNA stores hereditary information. Friedrich Miescher developed the technique for isolating DNA, but like many of his colleagues he did not believe DNA was a large or complicated enough molecule to contain any significant information (Dahm, 2008). For many years a common view in the scientific community was that proteins were the only viable candidate; researchers believed they were large, diverse, and complicated enough to be up to the task. It took years of research to discover the true nature of how hereditary information is stored and replicated.

An early breakthrough in our understanding of where genes are stored stemmed from the 1918 flu epidemic. English microbiologist Frederick Griffith examined two strains of Streptococcus pneumoniae, specifically the R and S strains (O'Connor, 2008). These were interesting to him because they varied widely in appearance and virulence. The highly virulent S strain had a smooth capsule, or outer coat, while the nonvirulent R strain was rough in appearance and did not have a capsule. Animal subjects exposed to the S strain died within days, but subjects exposed to the R strain survived.

Griffith was surprised to see that animal subjects exposed to a mixture of dead S strain and live R strain died, even though animals exposed to either the dead S strain or live R strain survived (O'Connor, 2008). When examining bacteria taken from the dead subjects, he noted that the formerly rough-looking R strain bacteria now had the smooth capsule characteristic of the S strain. He hypothesized that a chemical component from the virulent S cells had transformed the R cells, although he did not know through what mechanism this was possible. He referred to this chemical component as the "transforming principle" (Griffith, 1928).

Oswald Avery at Rockefeller University took note of Griffith's research and undertook a series of experiments to explain the transforming principle. Instead of culturing bacteria in animal subjects, he cultured samples in petri dishes, which gave him much better control over the experiments (O'Connor, 2008). He used a process of elimination to determine what component was responsible for the transforming principle, finding that R strain could not pick up traits from S strain if DNA was not present. He published his findings in 1944, arguing that hereditary information must therefore be stored in DNA, not in proteins, as was commonly believed. This was the work that inspired Erwin Chargaff to pursue his research into the structure of DNA wherein he disproved the tetranucleotide theory and set out his rules for base pairs (Aldridge, 2003).

This theory that DNA was responsible for storing hereditary data was not widely accepted until a series of experiments published by Alfred Hershey and Martha Chase in 1952 (O'Connor, 2008). They came up with a clever application of radiation signatures to track both the DNA and protein portion of bacteriophages infecting E. coli bacterium. This way, they could see whether it was primarily DNA or protein molecules entering and reproducing within an infected bacteria cell. The experiment showed that the majority of material entering the cell was indeed DNA molecules. Interestingly, they stated that more research was required, and that while DNA had some function in the process, they could not explicitly conclude that DNA was the hereditary material. However, the following year, Watson and Crick determined the structure of DNA, and this discovery also supported the function of DNA. Thus, the idea that proteins were hereditary material was finally laid to rest.

Junk DNA and the Known-Unknown

While our understanding of DNA is more advanced than it has ever been before, there is still a great deal that we are unaware of. On this subject, the question of what is DNA becomes mirkier. One such area is that of jumping genes, which scientists for decades disregarded as junk DNA that had no function and was merely biological noise of sorts. It will come as no surprise for those who are paying attention that we are beginning to see that jumping genes do have an important function, albeit one that we do not yet have a full understanding of (Pray, 2008).

In the 1940s, the geneticist Barbara McClintock discovered what she referred to as transposable elements (TEs) which were sequences of DNA that move from one location in the genome to another (Pray, 2008). For decades after this initial research, other scientists dismissed the relevance of TEs. McClintock, however, was among the first to suggest they might play some regulatory role in deter-

mining which genes are turned on and when this activation takes place.
Around the same time scientists Roy Britten and Eric Davidson speculated that jumping genes might have a role in different cell types and different biological structures. They hypothesized this might relate to how multicellular organisms have different cell types even though most cells share exactly the same DNA (Pray, 2008).

More recently researchers have been looking at TEs, and some believe jumping genes may account for as much as 40% of the human genome. It is also becoming a common belief among researchers that TEs may indeed have a regulatory function, as speculated by McClintock, Britten and Davidson (Smit, 1999). Recent research from Ogiwara et. al. (2002) has shown that TEs are sometimes shared between species with common ancestors. Among one such group, the oldest surviving species dates back 544 to 510 million years ago. It begs the question, why would the same sequences of junk DNA remain so well preserved among multiple species, and furthermore still be observed to jump from one portion of the genome to another if they were useless? This encourages speculation and is driving continued research into exactly what functionality TEs have.

Summary

DNA is the essential building block of life, and our understanding of it has undergone a dramatic evolution over the last century and a half. From a humble start as the mysterious precipitate first isolated by Miescher in his lab in Germany in 1869 (Dahm, 2008), DNA has become the focus of intense scrutiny from the scientific community and the general public for many decades. The discoveries listed in this chapter are far from an exhaustive list, but they offer a reasonable picture of the years of study that went into cracking open this mystery. The mystery of our shared heritage and how our bodies store and pass on genetic information.

These discoveries were not without their setbacks. Miescher's belief that proteins, not DNA, contained genetic information, was a major one (Dahm, 2008). For Levene, it was his hypothesis of a tetranucleotide structure (Pray, 2008). Watson and Crick struggled with where the phosphate fit in the structure of DNA (Aldridge, 2003). But with setbacks also came discoveries, great leaps in understanding that opened the door to further research, new ideas and possibilities. There is no one person who can be cited when answering the question of who discovered DNA. Many will colloquially point to Watson and Crick (Dahm, 2008), but the reality is far more complicated than a single research team. What Watson and Crick discovered was undoubtedly important, but it would not have been possible without the efforts of Franklin, Wilkins, Avery, Chargaff, Levene, Miescher and countless others.

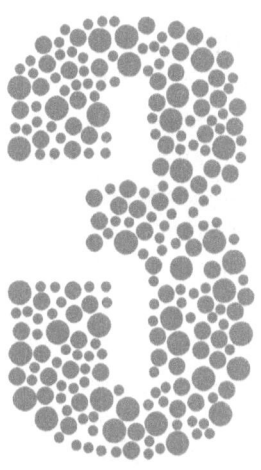

How Did the Discovery of DNA Impact Genetics?
Keshikaa Suthaaharan

Today, genetics is a prominent field broadly classified as the study of heredity (World Health Organization, 2002). The study of DNA has undoubtedly allowed the field of genetics to emerge in its present form. However, though DNA was initially identified by Friedrich Miescher in 1869, it was not regarded as the genetic material and Miescher's work on DNA was largely forgotten until the mid-twentieth century (Dahm, 2005). The discovery of DNA's structure and thus function in 1953, on the other hand, led to rapid developments in the field of genetics. As a result, for the purposes of this chapter, I will be examining the impacts of the discovery of the macromolecule's function rather than the impacts of the initial discovery of DNA. This chapter begins with a brief account of genetics until the discovery of the structure of DNA, and the remaining part of the chapter will discuss the influence of the discovery on the field of genetics.

As described in Chapter 1, the field of genetics has an interesting past in the history of humankind. The field of genetics is intricately linked with the study of evolution, with the theories of evolutionary theorist Charles Darwin. In particular, Darwin spoke about the concept of inheritance of traits between generations, which was the driving force of natural selection (Charlesworth & Charlesworth, 2009, p.758). Darwin, however, was unable to propose a mechanism for this inheritance (Charlesworth & Charlesworth, 2009, p.758). Gregor Mendel filled in this gap by describing three principles for inheritance between generations (Miko, 2009). He further described "factors of inheritance" passed from parents to offspring that would determine the offspring's traits (Sorsby, 1965). In 1909, the term "gene" was coined by Wilhelm Johannsen to describe these unknown heritable factors (Forsdyke, 2010).

After groundbreaking research was conducted by various scientists, as described in Chapter 1, it was determined that chromosomes contained genes. However, there was a debate about whether the DNA or proteins within the chromosomes were the heritable factors that had been described by earlier researchers. This debate endured over a large part of the twentieth century and was finally settled upon the structural elucidation of DNA (Pray, 2008). In 1953, utilizing the X-ray crystallography work of Rosalind Franklin, James Watson and Francis Crick determined the three-dimensional double helix structure of DNA (Pray, 2008; Watson & Crick, 1953a). This led to the elucidation of DNA replication and the genetic code, to technologies such as sequencing and polymerase chain reaction, and much more. Ultimately, the discovery of DNA structure and function paved the way for the development of the field of genetics into the form that we know today.

Exploring how DNA Replicates

Within the double helix structure, the nucleotide adenine is always paired with the nucleotide thymine, while the nucleotide cytosine is always paired with the nucleotide thymine, thus highlighting the specificity of nucleotide base pairing. Notably, one of the lines in Watson and Crick's renowned paper was "It has not escaped our notice that the specific pairing [of the nucleotides] we have postulated immediately suggests a possible copying mechanism for the genetic material" (Watson and Crick, 1953a). Indeed, the structure led to speculations about how genetic information could be replicated and thus transferred between generations (58).

Building upon the sentence in that first paper, Watson and Crick wrote a paper entitled "Genetical Implications of the Structure of Deoxyribonucleic Acid." In this paper, they discuss that even if only one strand was known, using the knowledge that adenine only pairs with thymine and cytosine only pairs with guanine, the second strand could be determined (Watson & Crick, 1953b). They hypothesized that this forms the basis for DNA replication: the hydrogen bonds break, each strand of the double helix acts as a template for replication, and two DNA molecules are produced at the end (Watson & Crick, 1953b). Watson and Crick's thoughts were mostly speculative and were met with some resistance in the scientific community. These doubts were eventually laid to rest five years later when Meselson and Stahl experimentally showed that DNA could indeed replicate in the manner predicted by Watson and Crick (Meselson & Stahl, 1958).

The discovery of the DNA structure inspired Arthur Kromberg, and he set out to determine the enzyme that could replicate DNA. In 1956, Kornberg described an enzyme isolated from Escherichia coli (a bacterial species) that could catalyze DNA synthesis (A. Kornberg et al., 1956). This enzyme was referred to as a "polymerase" in later papers (Lehman, Bessman, et al., 1958). These studies included the addition of E. coli DNA, since Kornberg assumed that the DNA would act as a "primer," in that it would act as a starting point for enzymes involved in elongating the chain (Friedberg, 2006). Contrary to this assumption, however, in 1958, Kornberg and colleagues determined that the DNA was instead acting as a template on which the polymerase could synthesize new DNA (Lehman, Zimmerman, et al., 1958). The same study also showed that the ratio of adenine residues was the same as the ratio of thymine residues, and the ratio of cytosine residues was the same as the ratio of guanine residues. These two results greatly supported the hypothesis of Watson and Crick about DNA replication, and brought additional experimental validity to the theoretical 1953 DNA structure.

These initial findings were quickly followed by the discoveries of the proteins involved in DNA replication. The replicating enzyme that Kornberg identified is now known as DNA polymerase I (Friedberg, 2006). In 1970, DNA polymerase II was discovered by one of Kornberg's three sons, Thomas Kornberg (T. Kornberg & Gefter, 1970). A year later, Thomas Kornberg identified DNA polymerase III, the enzyme that actually replicates E. coli DNA (T. Kornberg & Gefter, 1970). A couple of years after this, proteins capable of unwinding the double-stranded DNA helix were identified (Abdel-Monem et al., 1976; Abdel-Monem & Hoffman-Berling, 1976). These unwinding proteins became known as helicases (Geider, 1978). Other proteins involved in DNA replication were also discovered by the early 1980s, such as DNA primases, DNA ligases, and much more (Huebscher, 1984). By 1990, the roles of many enzymes and proteins involved in eukaryotic DNA replication had been elucidated: 1) DNA helicases separate the two strands of the double helix to produce two template strands, 2) Ribonucleic acid (RNA) primers are added by DNA primase to initiate replication, 3) DNA polymerases synthesize two new DNA strands on both template strands guided by base-pairing rules, 4) Topoisomerases relieve torsional stress that occurs during replication, 5) DNA ligase joins DNA pieces together, 6) Ribonuclease H removes the RNA primers, and 7) Auxiliary proteins play roles such as recognizing primers and coordinating the activity of DNA polymerases (Thömmes & Hübscher, 1990). For more information about DNA replication, see Chapter 4.

With the discovery of the structure of DNA, Watson and Crick had illuminated a potential mechanism for DNA replication; in the process, they predicted how DNA could allow for the transfer of genetic information between generations. Watson and Crick's discovery attracted the attention of Arthur Kornberg, who spearheaded the mission to determine the enzyme involved in DNA replication. The discovery of the DNA structure paved the way for a better understanding of DNA replication, a fundamental concept within genetics. Ultimately, the discovery of the double helix reinforced the notion that it was the molecular unit of heredity.

Cracking the Genetic Code
On July 8th, 1953, theoretical physicist and astronomer George Gamow excitedly addressed a letter to Watson and Crick after reading their article on the structure of DNA (Nanjundiah, 2004). In this letter, he excitedly commented on the discovery and stated that the discovery brought biology into the field of 'exact sciences'; in other words, it allowed biology to intersect with physical sciences (Rich, 1997). Later that year, Gamow wrote to Linus Pauling about his thoughts on the Watson and Crick model. He wrote, "Ever since I read the article of Watson and Crick last June, I was trying to figure out how a long number written in a four digital system (i.e. nucleic acid molecule) can determine (uniquely) a correspondingly long word based on 20-letter-alphabet (i.e. an enzyme molecule)" (Gamow, 1953). With these words, Gamow had proposed the concept of a 'genetic code,' a code that would relate the genetic information contained within DNA to the amino acid sequence of proteins (Nanjundiah, 2004; Nirenberg, 2004).

Inspired by the discovery of the DNA structure, a frantic race to decipher the genetic code began in the years following 1953 (Nanjundiah, 2004; Nirenberg, 2004). A major issue that emerged early on was that there were only four types of nitrogenous bases, but there were 20 amino acids that comprised proteins (Crick et al., 1957). The unit of genetic information that would encode an amino acid was known as a "codon" (Nanjundiah, 2004). Mathematically, this meant that at least a triplet of bases had to represent a single amino acid. However, with triplets of four different bases, there are 64 unique triplets, much more than 20. This discrepancy, dubbed the "coding problem" by Francis Crick (Crick et al., 1957), became a major preoccupation of the coders.

Gamow himself came up with the Diamond Code to explain the coding problem. In a letter to Linus Pauling, he described a lock-and-key mechanism in which amino acids would fit into "diamond" grooves formed by DNA (Gamow, 1953). The Diamond Code code proposed triplet codons, consisting of three nucleotides, that were ultimately equivalent to 20 amino acids (Nanjundiah, 2004). An attractive feature of this code was that a single nucleotide was involved in two consecutive triplets, leading to a one-to-one ratio of triplet codons and amino acids in the chain (Nanjundiah, 2004). Ultimately, however, the Diamond Code was discounted when another researcher named Sydney Brenner showed that it imposed constraints on proteins that were not realistic (Brenner, 1957). Even though his code was proven incorrect, Gamow showed how a biochemical problem could be reduced into an abstract theory of coding, which later proved significant (Crick, 1966).

In 1953, in order to liven up the race, Gamow began the "RNA Tie Club" (Crick, 1966). This group featured notable researchers, including Francis Crick. Many RNA Tie Club codes were focused on determining how DNA could be used directly as a template for protein synthesis (Crick, 1966). There was also a large emphasis on the idea of exactly 20 codons to match the number of proteinogenic amino acids (Goldstein, 2018). Similar to Gamow, other members, including Richard Feynman and Edward Taylor, proposed overlapping codes (Nanjundiah, 2004).

By the late 1950s, there was evidence that ribonucleic acid (RNA) was the template of protein synthesis, not DNA (Woese, 1969). Accordingly, Crick took a novel approach from other members of the RNA Tie Club. Firstly, Crick proposed his adaptor hypothesis in a letter to Gamow: that an entity acted as an "adaptor" between the DNA template and the protein product (Crick, 1955). Though his theory was not entirely correct, Crick had identified a missing piece of the coding problem and a source of failure within the RNA Tie Club: that DNA was not the direct template for protein synthesis. Based on this, Crick proposed a triplet code that was based on non-overlapping codons. A major premise of this code was that if two codons that produced amino acids were placed side by side, any overlapping triplets between the two would not encode amino acids (Crick et al., 1957). In other words, adaptors only existed for a subset of all possible codons (Biro, 2008). Based on mathematical analysis, it was determined that the comma-free code would produce 20 codons — the magic number (Crick et al., 1957). This was an especially appealing code to members of the scientific community. As it turned out, however, no members of the RNA Tie Club would actually solve the coding problem.

Despite this, the concept of mRNA would be key in the race to solve the coding problem, which came to a dramatic finish a mere 5 years later after Crick's proposal of the comma-free code (Crick, 1966). Independently of the RNA Tie Club, Marshall Nirenberg had been quietly working to crack the code. Nirenberg was the first to show that messenger RNA (mRNA) was indeed the intermediate between DNA and protein, as only RNA would stimulate incorporation of amino acids into proteins (Nirenberg, 2004). In 1961, Nirenberg discovered that the RNA sequence UUU corresponded to only phenylalanine (Matthaei & Nirenberg, 1961; Nirenberg & Matthaei, 1961). By 1965, Nirenberg had identified which codons correspond to which amino acids (Brimacombe et al., 1965). The complete genetic code, unlike many of the codes proposed by RNA Tie Club members, did not consist of 20 codons. In fact, the genetic code was degenerate, so one amino acid could be encoded by more than codon. There were a total of 64 codons that could encode for the 20 proteinogenic amino acids. Less than fifteen years after Watson and Crick's monumental discovery of the double helix structure, the genetic code was cracked.

As described here, the discovery of the DNA structure was the spark of the coding race of the 1950s and early 1960s. It caught the interest of George Gamow, who made significant contributions to popularizing the coding problem. Due in large part to Gamow and his RNA tie Club, the coding problem attracted the attention of numerous researchers. Though the RNA Tie Club did not succeed in its original purpose, Gamow's enthusiasm and the efforts of the club members galvanized the field of genetics (Nanjundiah, 2004). In many ways, Gamow created a turning point in the field of biology, leading to immense interest within the then infantile field of genetics and providing the basis for numerous discoveries. Furthermore, important hypotheses emerged from the RNA club that were significant to genetics, such as Crick's adaptor hypothesis. Eventually, this flurry of interest led to the discovery of mRNA's role as an intermediary in protein synthesis and the elucidation of the genetic code. Ultimately, over the 1950s and 1960s, the field of genetics rapidly began to gain momentum.

Developing DNA-Based Technologies

The discovery of the structure of DNA, as well as its implications for DNA as a genetic material, illuminated numerous possibilities for the field of genetics. In this way, it provided the necessary basis for technological advances within the field of genetics. In particular, in the years following 1953, exciting developments in recombinant technologies, sequencing technologies, and the polymerase-chain reaction (PCR) occurred.

Recombinant technology is based on modifying DNA, but specific tools were needed to accomplish this. Early on, Werner Arber hypothesized that individual bacterial strains could produce an enzyme that could recognize and cleave foreign DNA at specific sites (Arber and Linn, 1969). This enzyme eventually became known as a "restriction enzyme" (Roberts, 2005). In 1968, Meselson and Yuan isolated a second restriction enzyme from one E. coli strain that could cleave foreign DNA from other E. coli strains (Meselson & Yuan, 1968). Three years later, restriction enzyme endonuclease R was shown to exhibit specificity, as it could cleave Simian Virus 40 DNA into eleven specific fragments (Danna & Nathans, 1971). This same study separated the digested fragments using gel electrophoresis, a method commonly used to separate components of a solution (Danna & Nathans, 1971). Importantly, this combination of gel electrophoresis and restriction enzymes was quite similar to practices used today (Roberts, 2005). Furthermore, endonuclease R (now known as EcoB) was the world's first Type II restriction enzyme, the type that is significant to recombinant technology as the ability to detect specific DNA sequences allows for directional cleavage (Roberts, 2005). Other Type II restriction enzymes were identified in the following years, including EcoRI (Roberts, 2005). From these early discoveries, Type II restriction enzymes were ingeniously used to produce modified (recombinant) DNA — in one such method, EcoRI was used to digest both a plasmid vector and a DNA fragment to generate compatible "sticky ends," and then an enzyme ligated the two together (Cohen et al., 1973). Recombinant DNA technology had officially been born.

Efforts to sequence DNA began in the late 1960s. Many early methods, such as utilizing the incorporation of radioactive nucleotides, were only effective in sequencing short stretches of DNA (Heather & Chain, 2016). In 1977, the major sequencing breakthrough originated from Alfred Sanger, who developed the chain-termination (or dideoxy) technique (Sanger et al., 1977). In this method, Sanger initiated a normal DNA replication reaction with a certain amount of the template DNA, all of the necessary enzymes and the four nucleotides. However, he also added dideoxynucleotides into the mixture. Dideoxynucleotides do not possess a hydroxyl group on the third carbon of the pentose ring, unlike their nucleotide counterparts. This hydroxyl group is required for polymerization via a DNA polymerase, so if a dideoxynucleotide was added to a growing DNA chain, replication would stop. With enough time, DNA fragments of all possible lengths could be produced. If four different trials with the four different dideoxynucleotides were conducted and the resulting fragments were separated on an electrophoresis gel, then the sequence could be inferred from the results (Sanger et al., 1977). First generation sequencing, as this method became known, was the first of many successful sequencing technologies (Heather & Chain, 2016). Newer technologies include methods based in pyrophosphate production, an example second generation sequencing, and methods based in nanopores, an example of third generation sequencing (Heather & Chain, 2016).

In the late 1980s, work was ongoing to develop methods of synthesizing large quantities of DNA. The idea of base-pairing, as described in Watson and Crick's structure, was highly important in this research as it conferred specificity between separated strands of a single double helix (Mullis et al., 1994). From sequencing, which had become a routine practice at that point in time, it was known that short blocks of DNA (oligonucleotides) could be used as "primers" to synthesize desired DNA sequences (Mullis et al., 1994). Building upon the principles of DNA replication, Kary Mullis developed the method of PCR in 1987 (Mullis & Faloona, 1987). This method utilized continuous rounds of DNA replication catalyzed by a DNA polymerase to generate multiple copies of a desired DNA sequence. Eventually, the thermostable Taq DNA polymerase was utilized in the PCR reaction to improve the effectiveness of the technique (Saiki et al., 1988). PCR quickly became an important tool within genetics, particularly in conjunction with sequencing and as part of a diagnostic tool (Kary B. Mullis et al., 1994).

As shown above, the discovery of the structure and role of DNA in replicating genetic information — coupled with information about DNA replication — eventually gave way to revolutionary technological advancements. These technologies would eventually become routine tools within the field of genetics, which highlights how rapidly the field is growing at this time.

A New Era of Genetics

All of these technological advancements, made possible by the discovery of DNA' structure and function, led to a new era of genetics. The following years would show discoveries that would never have been imagined at the time of Watson and Crick's discovery. A couple of examples include the Human Genome Project, phylogenetic trees, and genetically modified organisms (GMOs).

The last quarter of the twentieth century marked great interest in sequencing genes, and eventually, entire genomes. During meetings arranged by the US Department of Energy, the idea of sequencing the entire human genome emerged (Palca, 1986). This initial idea steadily gave rise to the Human Genome Project (HGP), an international collaboration that aimed to uncover the sequence of the entire human genome (Lander et al., 2001). A physical and genetic gap of the human genome was first produced. In addition, prior to sequencing the human genome, strategies were tested on the yeast and worm genomes. These studies proved successful, and led to the development of the two-phase strategy implemented in the HGP (Lander et al., 2001). This project ultimately made use of recombinant, sequencing and PCR technologies (Heather & Chain, 2016). In the first phase, the "shotgun" phase, the human genome was divided into fragments and then further divided into overlapping fragments that were then sequenced (Chial, 2008). These were then connected together to form a continuous sequence. Phase itself led to 90% of the entire human genome. Phase 2 involved filling in any gaps from Phase 1 and resolving any ambiguous areas (Chial, 2008). Due to the development of simultaneous sequencing technologies and other advancements, the HGP completed Phase 1 in 2001, much earlier than planned (Heather & Chain, 2016; Lander et al., 2001).

Recombinant DNA technologies led to genetic engineering, the ability to change DNA within genomes. From this, came the insertion of new genes into organisms to confer new properties. In agricultural plants, this was commonly done to increase crop yield and generate herbicide resistance (Phillips, 2008). A notable example of a modified crop was Bt corn, which was genetically modified to naturally produce proteins that are toxic to insects, thereby making the crop resistant to insect infestation (Hellmich, 2012). Later efforts developed technologies for genome editing, the process of selectively modifying desired sequences of DNA (Li et al., 2020). One such technology, CRISPR-Cas9, will be discussed in the next chapter.

The development of DNA sequencing led to major developments within phylogenetics. During the 1980s, DNA-based phylogenetics became more and more common (Brown, 2002). In particular, once different DNA sequences from different organisms were sequenced, they could be aligned with each other. Based on this alignment, and calculating the level of homology between the sequences, a phylogenetic tree — a diagram showing the evolutionary relationships between organisms — could be constructed (Brown, 2002). These phylogenetic trees are useful for studying evolution, and for organizing information about biodiversity (Baum, 2008).

The last quarter of the twentieth century ushered in a new era of genetics, one focused on understanding the human genome, and even modifying sections of genomes. The stage was set for new possibilities in human health, such as examining the human genome for mutations and treating diseases by modifying genomes. At the heart of all of these groundbreaking discoveries is Watson and Crick's discovery in 1953, and the research that preceded it.

Summary
The discovery of DNA's structure and function was the breakthrough that revolutionized the field of genetics and acted as the foundation for the countless

discoveries that followed. The 1953 discovery directly led to a detailed understanding of how DNA replicates. Interestingly, the discovery even attracted a physicist named George Gamow who attracted countless researchers to the field by initiating the great coding race. Furthermore, the discovery provided the basis for groundbreaking technologies such as restriction enzymes, sequencing and the polymerase chain reaction.

Today, the DNA technologies developed during the latter half of the twentieth century have found their place in clinical practice. As will be discussed in Chapter 4, they have found utility in understanding the heredity of diseases, diagnosing and screening diseases, and much more. In many ways, the 1953 discovery popularized the field of genetics and helped it grow into the form it is today. Ultimately, Watson and Crick's discovery had a revolutionary and lasting impact on the field of genetics.

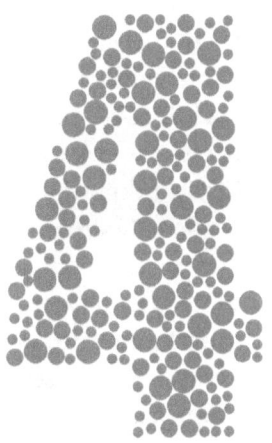

Why is DNA Worth Investigating?
Paawan Virdi

It is clear that humanity has come a long way from our early ideas of inheritance. We now know that DNA is the hereditary material, but why should we care about it beyond that? An understanding of the structure and function of DNA has opened up many possibilities in clinical settings, scientific research, and beyond! For instance, the understanding of genetics provided by DNA research has enabled the field of personalized medicine, revolutionizing the way that patients are perceived and treated. Understanding the processes that DNA undergoes has revolutionized the field of biotechnology. Genetics and DNA have also provided us with new philosophical perspectives, as exemplified by the perennial question of nature versus nurture. It is clear that DNA research is an interdisciplinary field with a scope encompassing many areas of knowledge. The discoveries we have made have had immense benefits in multiple fields of science and to humanity as a whole.

Our Own Code: DNA Profiling and Genetic Screening

DNA is known as the substance that makes us all unique. Your DNA and my DNA are very different, and our DNA has played immense roles in determining how we have developed. After all, almost no two individuals have identical sets of DNA. Even identical twins tend to differ by a few base pairs (Jonsson et al., 2021). Because of this wonderful uniqueness, our genetic code can be used to distinguish us from the billions of other humans on the planet. The fact that our DNA can act like fingerprints, and the ability to conduct genetic analysis has plenty of applications.

One of the ways that DNA can be analyzed is through gel electrophoresis. It is a method of identifying DNA fingerprints without even sequencing the genome. That is, it is a method of comparing different genetic data without looking into the exact genetic code. It is quite a clever technique! It utilizes a type of protein known as restriction enzymes. Most restriction enzymes look for specific palindromic sequences on DNA. A palindrome is a word that is the same backwards as it is forwards, such as the word 'kayak'. Palindromic sequences are those whose complementary sequence is the same as the original, but backwards. An example of a palindromic sequence is AACGTT, for which a complementary sequence for this is TTGCAA. Given the immense number of base pairs that would be present in a sample of DNA, it is likely that such a sequence occurs somewhere in the DNA. The trick here is, DNA from different individuals would have such sequences at different locations along the sample. This means that the DNA is cut into different sizes, as the number of base pairs between such sequences would vary. Now we have a sample of DNA that is cut into different sizes, so what? The different sized samples of DNA are separated using gel electrophoresis. As you can guess from the name of the procedure, the DNA is placed into wells that are made at one end of a conductive gel. A current is then applied that polarizes the gel such that the side with the DNA samples is the negative side, and the opposite side is positive. DNA is negatively charged as a result of the phosphate groups in its backbone, this means that there is an electrical force that pushes the DNA away from the wells towards the positive electrode. There are many factors that influence how fast the DNA fragments travel down the well, such as properties of the gel, the size of the DNA, and the magnitude of the voltage applied across the gel (Lee et al., 2012). Since the only thing that differs between the different samples is the size of the fragments, different patterns of DNA travel are observed between the samples. Since everyone's DNA is different, their DNA samples fragment differently, leading to different patterns that are observed in the gel. This allows for the comparison of DNA samples, and has applications in things like paternity tests and forensics.

Just like fingerprint comparison is used in forensics, DNA profiling is used to reference genetic material left at a crime scene with that of known suspects. The DNA sample can be placed in wells adjacent to those collected from suspects of a crime. As everyone's sample would yield unique bands on the gel, if the pattern obtained from the sample at the crime scene matches up with one of the suspects, it would suggest that suspect is the criminal. While the process is incredibly accurate, there is still some potential for error that can result in false incrimination ("DNA Profiling," 2019).

This is just one example of how an understanding of the structure and properties of DNA have been applied, and this is without actually sequencing the DNA itself. One example of how more precise methods of genetic analysis that involve determining the sequence of DNA is determining the presence of abnormalities in DNA that could be linked to genetic disease.

The Clinical Applications of DNA Research

The study of DNA has also had immense implications in clinical settings. Examples of this include the ability to tackle medical conditions that involve the processes of DNA: replication, transcription, and translation. An understanding of the hereditary material yields an understanding of inheritance. This is of importance when considering that genetic disease is heritable. In addition, understanding that every patient is literally coded differently leads to new perspectives and treatment methods that are optimized for each unique individual. This has led to the field of personalized medicine. Our understanding of our genotype, our phenotype, and everything in between has pushed the boundaries for conditions we can treat.

It can be said that every single disease has a genetic component. This is because certain individuals will inevitably be more prone to certain diseases than others (Genetic Alliance & District of Columbia Department of Health, 2010). Diseases that are direct results of anomalies in one's DNA are referred to as genetic disease. Genetic diseases have patterns of inheritance that are based on multiple factors. For example, one pattern of inheritance is autosomal dominant. This means that the gene corresponding to the genetic disease is inherited via one of the non-sex chromosomes. This means that gender does not play a major role in inheriting the disease. Usually, if we have some sort of mutation that causes an error in our DNA, it is compensated by the fact that we have two copies of each "homologous" chromosome, meaning that the mutated gene can be made up for by a functional gene on the other chromosome. In the case of dominant inheritance patterns, even a single copy of a defective gene can have consequences. An example of this type of genetic disease is Huntington's disease, a slow-acting disease that causes the degradation of neurons in the brain. As there is an equal chance for either allele to passed on, an individual with Huntington's disease has a 50% chance to pass on the defective allele—and thus the condition—to their progeny. The knowledge of inheritance patterns can be used to determine the probability that a child will be born with certain genetic conditions. The family history with a certain disease can be represented by pedigree charts, which identifies family members who were or are carriers of the mutated genes. This can be useful knowledge in counselling for families concerned about their children being affected by genetic disease (Brock et al., 2010). Genetic testing is often done to preemptively detect genetic or chromosomal conditions prior to birth (Public Health Agency of Canada, 2013).

Some diseases are not heritable in the same way, but are tightly involved with the processes that DNA undergoes. As a result, understanding the mechanism of these processes can prove to be of significant importance. For instance, treatment measures for diseases like cancer and HIV often involve tackling the processes of DNA replication and transcription respectively. DNA replication is the process by which strands of DNA copy themselves to produce more DNA strands; this process was explained in an earlier chapter. Transcription is the next process that follows in the procedure that utilizes genetic information. DNA is double stranded and is stored in the nucleus. The information from DNA must be sent to the ribosomes, which is the machinery in the cell that converts genetic information to protein. Unfortunately, double stranded DNA cannot easily leave the pores of the nucleus and travel through the cell to reach the cytoplasm easily. Instead, DNA must be transcribed into single stranded mRNA, much shorter strands that only contain the relevant coding information. This process is carried out by an enzyme known as RNA polymerase, which is responsible for separating the DNA strands and bringing together ribonucleotides to create an RNA strand complementary to a template DNA strand. After some protective modifications, this mRNA can then venture out of the cell to be used by the ribosome to produce the proteins our body's cells need (Clancy, 2008).

Understanding this process is one of the ways that we can combat complicated retroviruses such as HIV. This is because retroviruses utilize an unusual enzyme known as reverse transcriptase. As you can guess from its name, this enzyme kind of performs transcription in reverse. Once the virus infects the cell, it creates retroviral double stranded DNA from viral genetic material, which is often RNA. This DNA then merges with the DNA of the host cell. This DNA is then transcribed and translated using the machinery of the host cell into viral proteins that facilitate the proliferation of the virus. One class of drugs that prevent this are known as nucleoside/nucleotide reverse transcriptase inhibitors (NRTIs). Nucleotides are the individual building blocks for nucleic acid. They are different from nucleosides as they have a phosphate group attached at the 5' end. Once they are activated by the body and reach the cell, NRTIs act as analogues to the building blocks of nucleic acids, and are recruited by reverse transcriptase to elongate the growing viral DNA strand. However, the difference between NRTIs and normal nucleic acid building blocks lies in a key element of their chemical structure. They are modified such that they are missing a hydroxyl group at the 3' carbon. This functional group is essential to the elongation of a nucleic acid strand (Patel & Zulfiqar, 2021). Without it, the strand cannot be synthesized, preventing the virus from creating the viral DNA it needs to turn the host cell into a virus-spreading-protein factory! However, the superior proofreading mechanisms of human polymerases prevent this drug from significantly affecting regular DNA replication. This is an example of how our understanding of both the structure and function of DNA have presented us with the ability to attack unique diseases.

Another condition that has been addressed using our knowledge of DNA is cancer. Cancer occurs when the regulatory mechanisms that limit cell growth go astray, causing cells to grow and divide uncontrollably leading to tumors. For a cell to divide, it needs to replicate its DNA such that each daughter cell has a copy of the genome inside it. This process is complicated, and there are many drugs that take advantage of this process, or processes that come after to prevent the proliferation of cancer cells. Examples of such drugs include alkylating agents, which result in DNA crosslinking. DNA crosslinking is when unintended covalent bonds form between two nucleotides of DNA, they can be on the same strand (intrastrand) or between the two strands (interstrand). The latter are the most lethal as they can prevent the separation of the strands which is necessary for DNA replication (D'Andrea, 2015). They also cause breaks in the DNA strands, errors in base pairing, and prevent cell division (Ralhan & Kaur, 2007). Ultimately, the result in the death of the cell, but are much more lethal to cancerous cells when compared to normal cells due to the rapid proliferation of cancer cells suggesting an increased reliance on DNA replication (D'Andrea, 2015). A detailed understanding of this process has enabled us to push the limits of medicine and combat these unusual conditions which our immune systems fail to overcome.

One recent example of how our understanding of DNA has pushed the limits of medicine is the recent development of mRNA vaccines. The SARS-CoV-2 Vaccine was the first of its kind to be approved for use. The vaccine consists of mRNA that is wrapped in a lipid based nanoparticle that is used as a delivery system. When the mRNA is delivered successfully into the cell, the cell's own machinery (the ribosomes) use the mRNA code to synthesize just the spike protein that is found on the SARS-CoV-2 virus (Pardi et al., 2018). The entire spike or some fragments are then displayed on the surface of the cell to act as antigens, letting the body know to prepare its immune system for when it encounters the spike protein again. That way, the body's immune system is already ready to carry out a thorough defensive response when we come in contact with SARS-CoV-2. By only using the mRNA and not other components of the virus which is used by other traditional vaccines, the risk of the viral DNA entering that of our cells is avoided (Pardi et al., 2018). Another interesting aspect of the mRNA vaccine is that it utilizes a modified version of uracil, which is one of the four nucleotides found in RNA. Instead of using uridine as the nitrogenous base, the modified uracil uses N-methylpseudouridine. The reason behind this is that there is an increase in its stability and its capability for translation, and also a decrease in its immunogenicity, i.e. the likelihood of an immune response (Pardi et al., 2018). The mRNA vaccine is the culmination of decades of research into nucleic acids, providing aid to millions of people around the world in one of the largest international crises in recent history.

It is clear that much of our uniqueness is literally coded into us by our DNA. When compounded with all the other factors that make us different, it is clear that no two patients will be completely identical. The concept of personalized medicine has emerged relatively recently, and involves customizing treatment methods to better fit the needs of the individual. It involves dividing up an overall diagnosis into multiple categories that differ with characteristics of the patient. For instance, a certain allele may influence how effectively an individual metabolizes a certain drug. Thus, for the same diagnosis, treatment methods may vary based on the alleles carried by an individual. Plenty of research has emerged that suggests that genetics, age, overall health, and many other factors play important roles in how an individual responds to a drug (Vogenberg et al., 2010). One example of how personalized medicine is applied is in the realm of cancer treatment. Genetic differences in the ERBB2 gene determine how much a protein known as HER2 is produced. When there is a lot of the protein produced, a drug called Herceptin can be prescribed for treatment, which is not usually prescribed when HER2 levels are low. In addition, individuals with this mutation who develop stomach cancer can also be treated with this same drug, demonstrating how genetic similarities guide similar treatment for different cancer (Canadian Cancer Society, n.d.)! The importance of genetic information in the treatment process is clear. The field is still developing to target even more groups of patients based on variety in genetic profile to increase the effectiveness of treatment, and provide new perspectives to the field of medicine.

DNA Research and Biotechnology

Advancements in our knowledge of nucleic acid structure and function has enabled the field of biotechnology. Understanding the processes undergone by the DNA of different life forms allow us to utilize it for applications ranging from drug development to food! For instance, our current supply of insulin is primarily produced through recombinant DNA technology (Baeshen et al., 2014). This involves primarily creating DNA complementary to the mRNA that is used to produce insulin. This is done using the reverse transcriptase enzyme that was mentioned earlier. It is pretty interesting to see how humanity has harnessed the abilities of troublesome viruses for our own benefit. As mentioned in Chapter 3, the DNA fragment can be introduced into the plasmid of E. coli using restriction enzymes and ligating enzymes. We now have a bacteria that possesses the DNA required to produce human insulin, transforming them into insulin-producing factories (Miller & Baxter, 1980)! This method has also been applied to produce many more therapeutic agents that humans need, like vaccines and hormones. However, the applications of this go far beyond just the fields of medicine and research. Industrial products can also be produced, such as enzymes that can be incorporated into laundry cleaning tools (Adrio & Demain, 2010). Genetically modified organisms can also be produced in order to increase nutrition and yield while decreasing cost (World Health Organization, n.d.). However, there are some debated concerns regarding the potential consequences of GMO use such as immune system reactions and the spread of herbicide resistance to unwanted crops such as weeds (Key et al., 2008). Genetically modifying the DNA of other organisms is definitely not the limit. With the development of systems like CRISPR-Cas9, we could begin modifying our own DNA to address mutations causing genetic disease (Liang et al., 2015). The future applications of this technology are discussed in later chapters.

DNA and Who We Are
Our understanding of DNA has also led to studies that teach us about ourselves and look to answer philosophical and existential questions. It has taught us a lot about how the organisms we see today have evolved from their ancestors. Genetic research revealed that we carry a small percentage of genes from neanderthals: what we call cavemen, telling us about our relatively recent ancestors (Garlinghouse, 2019). However, if we expand our scope beyond just our closer relatives, we see a lot of similarities with all life. The near universality of the genetic code suggests that it may have evolved early alongside amino acid biosynthesis. It also supports the notion that all current life may have evolved from one common ancestor. Genetic analysis also allows us to illustrate the tree of life. When using genes as a basis for relating species as opposed to other characteristics, the method of classification is called phylogenetic classification. This form of classification utilizes nucleotide sequences of indicative genes to determine where an organism lies along the timeline of evolution and its relationship to other groups of organisms (Doolittle, 1999).

In addition to pushing the boundaries in discovering where we came from and how we came about, genetics also plays a big role in our understanding of who we are as individuals. The question of nature versus nature is one that has been popular in philosophical discussion for millenia. The ancient Greek philosophers Plato and Aristotle had differing opposing views on this, with the former stating that we are largely shaped at our moment of birth and the latter stating that we are molded by our experiences. Generations of philosophers have argued both sides of this debate, but the essence of the playing field has changed drastically thanks to our understanding of DNA and the role it plays in shaping behaviour and personality. There is more information on this in Chapter 11.

Summary

It would be impossible to cover all the various applications that stem from an understanding of DNA structure and function in a single chapter. It is clear that the field of nucleic acid research has been very fruitful, with implementations across many disciplines and industries. By understanding DNA, we begin to understand the fundamental processes that govern the development of life. In doing so, we are able to utilize these processes to facilitate societal development, from catching criminals, to creating biological drug farms, to enriching the nutrition and yield of the crops that we grow. The possibilities continue to grow as new technologies are developed that push the limits of what we do with DNA. We can only imagine how DNA research will progress in a year, and the implications it will have on the countless fields it influences.

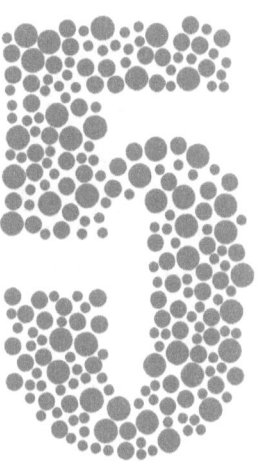

What Science is Involved in Studying DNA?
PRIYANSHU MAHEY

The study of DNA is a complex topic which builds upon eons of scientific progress. The difficulty of studying DNA comes with its particularly miniscule size as well as the length of individual DNA strands. Furthermore, the complex interactions in play within a DNA molecule have required years of research to figure out and are still not yet fully uncovered. The study of DNA is a field that still continues to grow to this day. This aforementioned field is a very interdisciplinary field, building the backs of many scientific fields including biochemistry, engineering, computer science, physics, and even more areas of expertise depending on the specific DNA study. The study of DNA itself is within the field of genetics, which primarily focuses on relating the genotype to phenotype. In order to actually study DNA, however, fields other than genetics will come into play.

Early Science Used to Study DNA
Early into the discovery of DNA, scientists had to focus more on the observations that they were able to make. In the late 1800s and early 1900s, there were no adequate tools to allow scientists to observe DNA. This meant that scientists had to rely heavily on observations of the phenotype and hypothesize how that would relate the genes of whatever they were studying. It wasn't until many years after DNA was hypothesized that adequate tools were available to properly study it. Many of the early scientific theories were based heavily off the pure scientific method. Many original studies behind genetics revolved around making observations of the appearance of an organism and relating that to the appearance or phenotype of the parents and/or siblings. This method relies on familial bonds to make predictions on genetic structures

and have helped us to develop better understanding of DNA. Let's go over some of the science of how DNA has been studied.

One very early insight study of DNA wasn't even focused centrally on DNA. In 1859, Darwin had published On the Origin of Species and here, not only did Darwin contribute immensely to evolutionary biology, he also made a huge leap forward in genetic understanding, describing genetic phenomena that are still relevant to this day (Liu et al. 2009). At the time, Darwin had no clue what DNA actually was but his observations on genetic inheritance and evolution were indicative of replicative ability of DNA. Darwin had written many hypotheses off of the observations he made during his voyages and they all relied heavily on him applying the scientific method and using logic to create theories. While not every scientific theory he made ended up being accurate, he was able to provide evidence for the fact that there must be something hereditary causing offspring to gain parental features. His studies have heavily impacted scientists for generations (Liu et al. 2018).

Alive at the same time period, Gregor Mendel also made huge strides in genetics relying solely on observation and the scientific method. Gregor Mendel's work on snow peas and progeny has left a huge impact and has helped us move our understanding on DNA and genetics forward. His study showed us a lot about how inheritance worked and again, just like Darwin, he had no clue what DNA was. Heredity is an important property of DNA and his studies pushed the field of genetics forward and helped to prove that there had to be some form of link that connected organisms, which we now know is DNA. He relied on nothing more than observations and his snow peas but yet was able to push the field forward immensely. Nowadays, there has been heavy debate when it comes his Mendel, his impact and his legitimacy have been put into question but it impossible to neglect the mark he has made on genetics research (Gliboff, 2015; Sandler, 2000)

So far we've discussed how important observations and theories have been in studying inheritance, progeny and the phenotype but we really have not yet discussed what DNA is. At the time of Gregor and Darwin, the mechanisms through which parents were able to pass down information was completely unknown. These are simply the earliest and simplest tools that we have to study DNA and its function. It took years to actually isolate DNA and even then scientists didn't know much about it. Friedrich Miescher was studying lymphocytes and by chance, one day while testing protease digestion, he came across a precipitate containing high amounts of sulfur and phosphorus. This precipitate, unbeknownst to him, was DNA and he had just invented an early method for DNA purification. His method relied on using alcohol, ether and alkaline solutions in order to purify and retrieve the first relatively clean sample of DNA (Dahm, 2007). Purification and separation of DNA are an important tool to study DNA and these concepts have enabled scientists to further deepen their research.

The Science Behind the Discovery of DNA's Structure

It took a while but as technology advanced, it enabled for better study of DNA. At this point in time, the science of DNA actually begins to focus more so on the structure and actual molecule of DNA rather than simply its role in phenotype. In 1952, Rosalind Franklin was the first to study the structure of DNA using X-ray crystallography, a technique still utilized today (Klug, 1968). X-ray crystallography is a tool which utilizes a beam of x-rays to decipher molecular structure of macromolecules. Upon hitting the molecule, the beam of x-rays disperses and their diffraction is measured and calculated in order to determine the 3D structure of whatever causes them to diffract (Smyth, Martin, 2000). This tool is still utilized today to learn more about structure and a common application of it is to see how changing a certain variable leads to alterations in the structure of a DNA strand such as in an experiment done by Egli et al. (2000). This tool itself was vital in helping to study DNA and is the means by which Watson and Crick were able to crack the code and figure out some of the most important details about DNA.

The Science Behind Modern DNA Studies

Modern studies have made excellent use of new and emerging technology. Technology advancements in biotech and our improved understanding of DNA has helped researchers to create novel methods to study the inner workings of DNA. The growth of the fields of bioinformatics and biotechnology has ushered in a new age of DNA studies. There are a lot of different tools we can now utilize to understand DNA from different perspectives. Let's go over some of the more interesting ones.

One giant leap in DNA studies is genome editing. The ability to edit the genome is significant as it allows scientists to create changes in the DNA and enables us to understand more how changes in DNA affect phenotype. Clustered Regularly Interspaced Short Palindromic Repeats (CRISPR Cas9) is one of the most widely used and available genome editing tools in modern day (National Institute of Health (NIHa), 2017). CRISPR Cas9 relies on a guide RNA to lead the Cas9 protein into the nucleus to make its cut at very specific locations. The Cas9 protein is found in many bacteria and serves an immunological function, but here in eukaryotic cells, it is applied to create cuts and make space for something called a repair strand. This repair strand is a strand of DNA that is made of the sequence we want in the genes and via the cell's own repair mechanisms, it will be inserted and used to repair the broken DNA. This allows for us to successfully replace DNA we don't want (Ma, 2014).

Gene editing is not the only tool that helps us to study DNA. Polymerase Chain Reaction (PCR) is a hugely helpful tool when it comes to studying DNA. PCR is a tool that replicates specific strands of DNA and allows us to study and make observations of the same copy of DNA. It works by first heating up DNA, which then causes the double stranded DNA to break apart and become separated strands. Then, RNA primers are attached and a specialized polymerase grows each chain of DNA. This then results in many duplicated DNA chains (NIHa, 2020). This tool is incredibly helpful in helping us to study DNA as it provides us ample copies of DNA to study (Alonso et al., 2004).

Another scientific tool important to the study of DNA is gel electrophoresis. Gel electrophoresis is a technique that applies electricity to separate macromolecule fragments based on their size. This mechanism allows a researcher to directly observe and study separated DNA. It works by applying a current which will affect the speed DNA travels proportionally to DNA molecules size. This DNA will then be put into a gel electrophoresis machine where an electric charge is applied. The DNA will travel through the gel in a distance that is proportional based on its charge, and this can be used to determine the size of the molecule (Lumpkin et al. 1985; Aaij, Borst, 1972).

Furthermore, another commonly used tool to study DNA is microarray analysis. Microarray analysis is a tool that is used to conduct large studies on the DNA of a population. This tool can be used to explore how often a gene is turned on in large sample sizes and can be applied towards clinical testing (NIH, 2020b). The way this works is by having the two strands of DNA denature and then having a single nucleotide strand complementary to the sequence we're searching for. The DNA strands and the strands complementary to it are then mixed together and this leads to binding which will be scanned and picked up by a specialized chip (Helen, 2002). This technique is especially effective for large scale genomic studies where we want to find which kinds of genes are prevalent in a population size.

These scientific tools are nowadays heavily utilized within genetic studies. These are highly useful as researchers are able to gain a lot of information. Another tool that is becoming more and more useful in studying DNA as time goes on is computers. This is because computers have the ability to store large amounts of data, work with the data and allow everyone can create their own scripts and programs. Researchers are able to easily conduct bioinformatic studies as well as use programming languages to work with large amounts of data relatively quickly. A huge feat that has benefited from computers is the Human Genome Project. One huge feat was the Human Genome Project. The human genome project sought to store and record all of the human genome and this arduous task would not have been possible without the storage power of modern computers and relied on computers to assemble the continuous stretches of DNA into the entire genome (NIH, 2020c).

Additionally, next generation sequencing is a term referring to a revolutionary field of DNA sequencing technology which relies heavily on the use of modern computing. It allows for any researcher to analyze the entire human genome and allows for researchers to conduct difficult studies on the genome. It has huge potential applications when it comes to healthcare and medicine (Behjati, Tarpey, 2013). Next generation sequencing utilizes computers in a productive way and enables for large scale DNA studies.

In general, there are many computer aided technologies that have advanced research or simply provided better universal access to researchers. The freedom of the internet has also allowed for researchers to be connected as now they can simply search up and share genomes with each other. This allows for the building of gene and protein databases which are easily accessible to anyone, enabling for more studies to be conducted. This enables for a wider availability of data for use and enables researchers to collaborate better (Iranbakhsh, Seyyedrezaei, 2011).

The Science Behind Future Studies of DNA
While there are many different ways to already study DNA, as technology advances, researchers continue to develop better and more unique ways to study DNA. We have a long way to go in order to fully understand how it works but as more technologies are developed, we will gain an even better understanding of DNA. There are many different types of technology being developed and a lot of them show great promise in expanding the field of DNA studies. One of the most prominent rising technologies are deep learning and artificial intelligence (AI) . Deep learning and AI enable computers to essentially think and solve problems that require complex thinking capabilities. They're also able to process large amounts of data quickly which, when processing DNA, is crucial. An example of what is being mentioned here is Alphafold, a deep learning tool which is able to predict the structure of a protein based on its initial amino acid sequence. This algorithm has been able to do what scientists never could — predict the structure of DNA (Senior et al., 2020). This challenge has stumped scientists for a while and it simply goes to show the promise AI and deep learning hold to solving some of our more complex

issues. While this is already being applied to DNA, AI and deep learning are still earlier in their lifecycle; however, they have already been applied in ways that we have never imagined many many years ago.

Another upcoming technology that shows a lot of promise is supercomputing. Supercomputers are immensely powerful and are able to do things in mere seconds that would take a regular computer days. While not readily available at the moment, the desire for them is rapidly growing as are the amount of research going into them. Supercomputers have the ability to run complex simulations and conduct large statistical research which will be incredibly beneficial in the aforementioned field of AI and deep learning (Abraham, 1986). In the past, researchers have benefited immensely from the growth of power in computers, enabling them to conduct better experiments (Nicholas et al., 1991).

Overall, as other fields such as computer science, physics, engineering, biochemistry grow, researchers are able to produce better studies. DNA studies are complex and the interplay from different areas means that studies rely on multifaceted approaches to improve our understanding of DNA. There are still a lot of things we do not know but with better tools we will continue to be able to learn more about the secrets of DNA.

The Multidisciplinary Study of DNA

It is incredibly apparent that the study of DNA has significantly grown and become an incredibly rich and diverse field from the point of it's discovery onwards. In this chapter alone, you have read about different methods that revolve around, biology, chemistry, physics, computer science, engineering, and pure analytical science. To study DNA, researchers need all these disciplines to help them uncover its secrets. To illustrate, let us look at the field of ancient DNA which looks at DNA samples from a long time ago. This field tells us a lot about ancient creatures such as dinosaurs, extinct animals or humans from a long time ago. This field requires another field we haven't discussed yet; archeology. Archeology expertise pieces together the historical context of the DNA, as well as another unique field, geology, to understand how the DNA was able to be preserved. It will also pull from histology, chemistry and molecular biology to be able to understand its composition. Furthermore, a lot of the tools engineered run on principles heavily rooted in physics, for instance, X-ray crystallography. Going forward, zoology, bioinformatics, botany can also be applied on a need by need basis. Then, computer science and engineering expertise are needed in order to develop the required tools (Parducci, 2019; Cipollaro et al., 2004). No matter what aspect of DNA researchers want to study, it will require a wide plethora of knowledge from many different scientific fields.

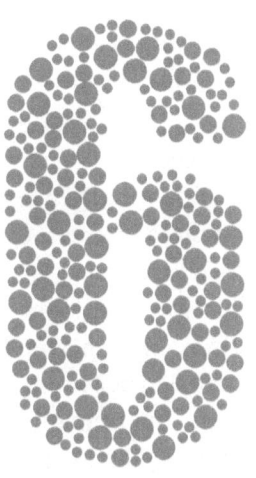

What Have We Learned About DNA in Recent Medical History?
GHULAM AISHA

Medical technology has significantly advanced over the years. Consequently, we have also come a long way in terms of what we know about DNA today and how it can help us provide the most effective medical treatments for diseases we once knew very little about. The original discovery of Watson and Crick helped form our basic understanding of phenotypic traits, DNA sequencing techniques and provided us with the ability to individually tailor treatments for patients by looking at virus and genetic mutations. Mutations refer to the "molecular alteration in the DNA sequence of a gene, with no inference made concerning the effect of this alteration on gene expression or function of the protein product" (Ooi, Gongska, Durie and Freeman, p. 2202, 2010). Effectively treating medical conditions saves time and money for both healthcare providers and patients. Genetic testing allows us to see if there is an increased risk or susceptibility for certain genetic conditions, allowing physicians to find them early on and stop them immediately. This chapter will discuss DNA-based personalized medicine, pharmacogenomics in patient care, genetic testing and gene therapy.

DNA-based Personalized Medicine

DNA-based personalized medicine is able to tailor a treatment specifically to an individual based on their unique genes. It relies on the principle that not all treatments will work for every patient. In essence, it is to tailor a drug to a person's genetic makeup inorder to provide the best possible response and lower the chances of any adverse reactions (Heid, 2015). Many people today are getting genetic testing done in order to know if their DNA points to an increased risk for genetic diseases such as Alzheimers, ovarian cancer and breast cancer. In particular, many places are already using DNA to tailor treatments specifically with regards to cancer treatment and cardiovascular disease treatment. For example, recently in St. Louis, researchers compared the gene sequencing in healthy tissues compared to diseased tissues in patients with advanced melanoma. They were able to pinpoint the unique mutations and create specific vaccines that increased the strength of a patient's T-cells (a type of immune cell) in fighting against the cancer (Carreno et al., 2015). Accordingly, similar advancements in the fight against neurological disorders such as epilepsy, infectious disease therapies and painkillers are expected to be found (Heid, 2015). However, there are some concerns regarding costs, funding and the administration of the program that still need to be answered before it becomes something that can be fully implemented.

Pharmacogenomics

Another important discovery that has been made is that of pharmacogenomics, which is also known as drug-gene testing. "Pharmacology" is the study of the uses and effects of medication on your body and "genomics" refers to the study of genomes (MayoClinic, 2021). It basically looks at the effect that changes or variants in your genes have on your body's response to medication (MayoClinic, 2021). This is similar to DNA-based personalized medicine, except in this case we specifically look at the impact genes have on drug effectiveness. This allows physicians to prescribe the most effective medications for medical conditions such as depression which have otherwise been difficult to combat (Heid, 2015). The current limitations for pharmacogenomics testing is that there is currently no testing for aspirin and several over-the-counter pain relievers, one test cannot help determine your response to all medications, and tests are not available for all medications (MayoClinic, 2021).

Genetic Testing

Genetic testing has quickly become popular when looking at the risk of carrying or passing on a number of genetic diseases in order to allow individuals to make more informed decisions about their health and bodies. Genetic test now play a role in a variety of clinical settings such as; pre-pregnancy and pregnancy counselling, newborn screening, cardiovascular screening, cancer screening, career screening, predictive screening, pharmacogenetics, neurological and developmental problems (i.e., X syndrome) and clotting and bleeding problems can all be explained through genetic testing which has only been made possible through medical advancements we see today (Blashki, Metcalfe and Emery, 2014; Ooi, Gonska, Durie and Freedman, 2010).

Biskup and Gasser (2012), detail the process on how genetic testing is done. Genetic testing is done by isolating DNA from blood leukocytes, cheek swabs, hair follicles or Guthrie spots. Usually, 10-20 ml of whole blood (EDTA) is taken and sent to a lab without first freezing or refrigerating it. After a period of 3-5 days, DNA is extracted. If the genes resulting in the neurogenetic disorder, molecular testing is done by looking at mutational analysis. Specific primers are used in the genomic area of interest and polymerase chain reaction (PCR) is amplified and the specific gene of interest is sequenced by Sanger sequencing. The process takes about four weeks and can cost between 100 and 3,000 euros dependent on gene size.

One common mutation includes point mutations where a few base pairs in the coding region are exchanged, deleted or inserted (Biskup and Gasser, 2012). This alters the amino acid sequence of the protein. If the mutation is one that is known to be pathogenic, it is easier to interpret (Biskup and Gasser, 2012). However, if there is a sequence variant that has not been described before, it can become difficult to determine the role of the sequence in the disease gene (Biskup and Gasser, 2012). A nonsense mutation terminates DNA translation and the loss of encoded protein can be identified as pathogenic in recessive diseases. However, missense mutations in dominant disease genes cause a change in the amino acid sequence that may or may not be harmful (Biskup and Gasser, 2012). Conserved amino acid sequences play a critical function and if the sequence is changed, it can be deleterious and lead to disease (Biskup and Gasser, 2012). In addition, if a mutation results in the disruption of the reading frame of a gene, alterations can result in the premature termination of transcription and may lead to a complete loss of the functional protein ("null mutations")" (Biskup and Gasser, p.1251, 2012). However, it must be noted that a loss of function does not necessarily result in disease. Over the years, scientists have found that the average human carries with them at least 50-100 potentially harmful mutations but these do not interfere with protein functioning (Biskup and Gasser, 2012).

Sickle cell disease
More and more people are resorting to taking part in genetic testing inorder to become more aware of the decisions they make about their health and bodies. For example, Ross (2015), found that due to medical advances women with sickle cell disease (SCD) are living into adulthood and are more likely to ask their partners to undergo genetic testing. This is because children with SCD give birth to children with SCD or those that possess a sickle cell trait (SCT) (Ross, 2015). This is dependent on their partner's genetic makeup. If their partner has SCD, the child has a 100% probability of being born with SCD whereas if they have

SCT the child will have a 50% probability of having SCT or SCD (Ross, 2015). If their partner does not have SCD or SCT, the child will only have SCT (Ross, 2015) This information plays a key role in the decisions women have to make about their bodies. Prior to the advancement in technology and knowledge about genetics, genetic testing prior to conception would never have been an option. Today, women are able to make informed decisions about their bodies and their prospective children by taking part in genetic testing.

Brugada Syndrome (BrS)

Brugada syndrome is a, "hereditary arrhythmic disorder that is a result of right ventricular conduction delay and ST-segment elevation in the right precordial leads and syncope" (Kaufman, p.1419, 2012). This results in sudden death due to ventricular fibrillation. This can only be detected under certain circumstances where an individual has fever, exposure to a sodium channel blocking drug and vagal stimulation. The clinical diagnosis occurs when physicians take into account personal history and family history of type 1 BrS ECG or sudden cardiac death (Kaufman, 2012). Patients are advised to take part in family screening alongside the recommended medical treatment. Family screening is primarily done through genetic testing because they are able to detect the disease causing mutation providing key information to both affected and unaffected family members. In this case, the SCN5A mutation was tested as being the disease causing culprit (Kaufman, 2012). However, genetic testing is not all inclusive and has its own limitations such as not accounting for disease causing variants (Kaufman, 2012).

Neurological diseases

Certain neurological diseases such as Huntington's disease are a result of the mutation of a single gene and a PCR-based assay would be able to determine that. PCR involves genes or genomic regions being "enriched by hybridization with complementary RNA baits ("gene panels"), and then sequenced at very low costs compared to conventional Sanger sequencing" (Biskup and Gasser, p.1252, 2012). However, diseases such as Alzheimer's, Parkinson's disease (PD), spastic paraplegias, epilepsies, ataxias, neuromuscular disorder, inherited neuropathies, dementias and amyotophic lateral sclerosis are a result of different mutations on different genes which makes it difficult to identify them even through genetic testing (Biskup and Gasser, 2012). Different mutations on different genes are called 'allelic' and 'genetic' heterogeneity and this makes it difficult to determine the phenotypic characteristics for the genotype. In addition," reduced penetrance and variable expressivity often does not allow one to define a simple one-to one relationship between mutation (or genetic variant, the "genotype") and disease ("phenotype")" (Biskup and Gasser, p.1249, 2012).

Genetic Therapy

Genetic therapy is just one of the many ways medical technology has advanced. This provides an avenue for solving genetic problems that may arise which could lead to long-term diseases. Gene therapy transfers genetic materials through a variety of techniques to help treat or prevent disease. Some of these techniques include replacing a mutated gene that is causing the disease with a healthy copy of the genes, introducing a new gene to help fight the disease and inactivating a mutated gene that is not functioning properly (Wilkinson and Borysiewicz, 1995). However, it should be noted that gene therapy is a promising treatment option in particular certain cancers, inherited disorder and viral infections, but it is still under study due to the risky nature of the treatment (Wilkinson and Borysiewicz, 1995).Thus, it is only being looked at for diseases that have no other cures. The expression systems for gene therapy require that genes target the appropriate cell, the delivery of the vector should be non-toxic to the target cells and scientists must determine the appropriate level, location and duration of expression in order to correct the disease phenotype (Wilkinson and Borysiewicz, 1995).

Nervous System

Genetic therapy targeted towards the nervous system can help treat the following conditions: lysosomal storage diseases, Alzheimer's disease and other amyloidopathies, Parkinson's disease, Spinal muscular atrophy, Amyotrophic lateral sclerosis and brain tumors (Bowers, Breakefield and Esteves, 2011). Genetic therapy inside the nervous system can occur through the central nervous system (CNS). However, it is enclosed by a blood-brain barrier (BBB) that helps regulate the chemical environment by preventing simple diffusion of small molecules and proteins into the bloodstream (Bowers, Breakefield and Esteves, 2011). As a result of this tight BBB, it becomes increasingly difficult for gene transfer vehicles from reaching the CNS (Bowers, Breakefield and Esteves, 2011). In order to use this route, a lot of gene therapy takes place by directly infusing gene transfer vectors into the brain to target the specific disease structure (Bowers, Breakefield and Esteves, 2011). Another route of entry for genetic therapy to take place is through the cerebrospinal fluid (CSF). However, research suggests that this method prevents the distribution of gene transfer vectors and disrupts normal CSF flow (Bowers, Breakefield and Esteves, 2011). The most ideal route of entry into the CNS is through vasculature. PEGylated immunoliposomes (PILs) deliver plasmids and RNAi to the brain or brain tumors (Bowers, Breakefield and Esteves, 2011). Non-viral nucleic acids are used to modify cellular processes inside the brain, however, one of the challenges is "target cell specificity and transient gene expression duration" (Bowers, Breakfield and Esteves, p.30, 2011).

Lysosomal storage Disease

Lysosomal storage diseases are childhood diseases that result from genetic deficiency in lysosomal enzymes within metabolic pathways, resulting in an accumulation of the substrate (Bowers, Breakefield and Esteves, 2011). This disease particularly does well under genetic therapy because it is a monogenic disease with an established genotype-phenotype correlations. It also allows the delivery of genes to the entire CNS (Bowers, Breakefield and Esteves, 2011). The two methods used in animal models that have proven to be successful are intraparenchymal infusion of recombinant viral vectors and bone marrow transplantation with ex vivo lentivirus vector modified-autologous HSCs (Bowers, Breakefield and Esteves, 2011).

Alzheimer's Disease (AD)

Alzheimer's disease (AD) is a neurodegenerative disorder that results in severe memory loss and cognitive impairments. Over the years attempts have been made to reduce the levels of the neurotoxic AB peptides or stopping the proteolytic release of AB from its amyloid precursor protein (APP) (Bowers, Breakefield and Esteves, 2011). This is done by using vector-based-therapies such as active and passive vaccines that target pathogenic AB peptides, delivery of small inhibitory RNA (RNAi), inhibiting expression of AB proteases and delivering cholesterol degrading enzymes into the brain (Bowers, Breakefield and Esteves, 2011).

Parkinson's Disease (PD)

Parkinson's disease (PD) is a neurodegenerative disease that impacts several regions of the brain, primarily impairing motor function due to a loss of the dopaminergic neurons in the substantia nigra in the basal ganglia (Bowers, Breakefield and Esteves, 2011). A number of factors play a role in getting PD such as environmental toxins and aging, but 16 gene loci are found in approximately 5% of the cases and hereditary risk was associated with 85,86 cases (Bowers, Breakefield and Esteves, 2011). Several different genetic therapies have been implemented to combat this disease. Namely the "upregulation of neurotrophic factors, generation of endogenous dopamine and alterations in neuronal circuitry, as well as implantation of supportive cells and dopaminergic neurons" (Bowers, Breakefield and Esteves, p.32, 2011). Gene delivery takes place through AAV vectors with CED injections into the striatum (Bowers, Breakefield and Esteves, 2011).

Brain Tumours

The most common brain tumor is called Malignant glioma tumors (GBM) and has very limited treatment options, with patients dying within 2 years of diagnosis (Bowers, Breakefield and Esteves, 2011). GBM is invasive and spreads quickly across the brain, reacting poorly to anti-angiogenic therapy (Bowers, Breakefield and Esteves, 2011). However, gene cell therapy shows some promise in treating this condition compared to traditional approaches. Traditionally, an effort was made to remove the tumor cells entirely, but gene therapy requires the removal of a bulk of the tumor and then injecting viral vectors or cells into the area in order to kill the remainder of the tumor cells (Bowers, Breakefield and Esteves, 2011). Suicide genes are inserted into the affected region in order to activate prodrugs that are able to pass the BBB and are able to convert into chemotherapeutic agents inside the tumor, allowing mutant forms of the virus to replication inside the tumor (Bowers, Breakefield and Esteves, 2011).

These are just some of the ways that genetic therapy is able to help alleviate those suffering from long-term illnesses.

Summary
We still have a long way to go in terms of enhancing our understanding of DNA and the role it can play in medical advancements. But we have achieved great feats already with our increased understanding of the role genetic testing and gene therapy play in the lives of individuals. As a result of advancements in genetic testing and the on-going research into gene therapy, individuals are able to make more informed decisions regarding their health and prospective children.

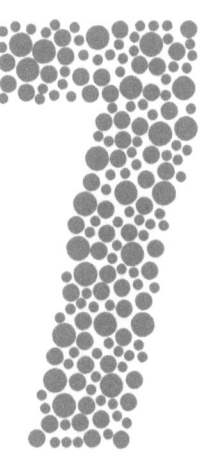

What Questions are We Still Asking About DNA?
NAWSHIN HAQ

Theories are the core of every scientific research community that is currently established in present day society. New hypotheses that build on existing theories are crucial for the constant flow of information output that is expected of these scientific associations. However, one of the most important aspects that initiate the development of these theories are questions. Without asking questions and being naturally inquisitive, humans would not have been able to build the literal scientific empires that we have today. One of the factors that humans often forget about scientific research, is that even long-established theories can be deconstructed with a single question. Questions, however, are expensive or rather the resulting experiment, that may occur as an aftermath of that question, is expensive. The governing boards of these research communities often do not want to fund these highly expensive projects for the fear of an insignificant result (Wood et. al., 2019). This refusal to continue delving into unchartered territory is echoed by most of the research-centered associations (Wood et. al., 2019). It is certainly simpler to keep looking at and studying the same ideas rather than trying to de-establish a theory or continue extensively studying the unknowns. However, studying the unknowns is a necessary evil that promotes an increase of knowledge and slowing or halting this process can have lethal effects.

Known Unknown: Protein Coding DNA

There was an explosion in the wealth of knowledge when protein-coding DNA was first discovered (Wood et. al., 2019). However, recently this rapid gain of information through scientific research has come to a halt (Wood et. al., 2019). Protein-coding DNA is no longer given the same level of importance as it was back in the 1990s, even in the most well studied species, such as the budding and fission yeast species (Wood et. al., 2019). In the past decade, the research for protein characterisation in budding yeast has only grown approximately 5% and about 4% for fission (Wood et. al., 2019). This is quite concerning since many uncharacterized proteins in the studied yeast organisms are preserved in the human body (Wood et. al., 2019). The researchers who conducted this analysis found that the science community has a bias towards characterizing proteins that are strongly intertwined with core biological processes (Wood et. al., 2019). However, they fail to study the preserved proteins that are correlated with environment dependent factors (Wood et. al., 2019). Although these proteins have strong implications for age-related diseases such as Alzheimer's, they are not given the same level of attention as the decade old studied characterized proteins (Wood et. al., 2019). The molecular-research associations still do not know what the appropriate functions for approximately 20% of these proteins are (Wood et. al., 2019).

The continuous theme displayed throughout this article when describing the research community, is the incessant pattern of remaining stagnant and not addressing core issues that may be plaguing many individuals, for fear of venturing into uncharacterized property (Wood et. al., 2019). This fact presents the multiple problems with remaining inert rather than continuously asking questions and trying to further one's research into an area that has yet to be explored. The questions we need to be asking, regarding conserved proteins are: what are the other characterizations for the remaining 20% of proteins (Wood et. al., 2019)? How are they associated with environmentally dependent disorders (Wood et. al., 2019)? Questions are extremely important to be asked by individuals in these communities and even more important to be addressed. The next topic (Non-coding DNA saga) will explore the pitfalls of failing to constantly forward inquiries and what may result as a consequence of not questioning a popularized theory.

Noncoding DNA: Human Genome Project and ENCODE Project
As students from elementary to high school we learn of DNA in a linear and redundant fashion. DNA is the basic building blocks of life because it is the hereditary material that builds our genetic code which gives instructions to build our proteins (MedlinePlus, 2021). In fact, this process of DNA to protein is so crucial that it is referred to as the central dogma of Molecular Biology (MedlinePlus, 2021). The process, as described in Chapter 4, is reiterated multiple times in an oversimplified manner and may make human beings susceptible to the common misconception that we have quite a bit of knowledge in regards to all DNA functionality. In reality, scientists have only

recently touched the tip of the metaphorical iceberg. Considering that scientists properly studied approximately 80% of proteins that are preserved in the human body, it is surprising to discover that protein-coding DNA makes up approximately 1-2% of the entire human genome (Wood et. al., 2019; Ling et. al., 2015). It is even more astonishing to discover that the rest 98-99% of the genome was unstudied for several decades after its discovery (MedlinePlus, 2021; Ling et. al., 2015).

The central dogma of Molecular Biology became extremely popularized after the discovery of DNA and thus protein coding areas of the human genome saw an extensive amount of study regarding biological processes and various diseases (Ling et. al., 2015). No scientist truly wanted to venture outside of this safe consortium of renowned study due to fear of an insignificant and unacknowledged result (Ling et. al., 2015). Inflicted with the fear of being ridiculed, no scientist wanted to study any areas of molecular research without a hyper focus on protein coding DNA (Ling et. al., 2015).

This scarce amount of research that was initially exerted onto non-coding DNA research led to the uniformed, perhaps cowardly, conclusion that this type of DNA was not worth studying (Ball, 2013). Thus, non-coding DNA was labelled 'junk DNA', and deemed unimportant to natural biological processes and diseases (Ohno, 1972). In 2003, when the results of the Human Genome Project (HGP) became known to the public, one of the major findings was that researchers did not know the function of nearly 3 million bases (nucleotide bases) of the human genome (Chial, 2008). This was largely because scientists attributed uselessness to the function of ncDNA and therefore scientists generally avoided studying this type of DNA (Ludwig, 2016).

The theory of ncDNA being futile or largely unimportant in function was not appropriately challenged until 2012 when The Encyclopedia of DNA Elements (ENCODE) project phase II was published (Dunham et. al., 2012). The ENCODE project is an ongoing, international research program that initially, during the pilot phase, studied protein-coding DNA and the function of these molecules (Dunham et. al., 2012). However, later the ENCODE project started studying function across the entire genome (Dunham et. al., 2012). The ENCODE project phase II, associated biochemical RNA and/chromatin function, to approximately 80% of the human genome outside of the protein-coding areas (Dunham et. al., 2012). This project threw many molecular biologists off (Ling et. al., 2015). There was now a frightening volume of unstudied materials, alluding to the functions of such a large, now important, percentage of the human genome (Ling et. al., 2015).

Noncoding DNA: Questions and Thoughts

Due to extensive efforts by multiple unrelenting scientists, recently there has been an increase in the amount of research conducted concerning ncDNA and, importantly, ncRNA.

Proteins determine the function of a given cell; therefore, each cell needs certain types and amounts of proteins to efficiently function (Nature Education, 2014). Gene expression refers to the theory that genes are regulated at each step of the DNA to protein process in order to manufacture the appropriate proteins for a given cell (Nature Education, 2014). This process of regulation begins during the transcription, after the DNA is transcribed to messenger RNA in the nucleus. During this process, the initial transcripts are modified: the introns (non-coding sections of the RNA transcript) are removed, and the exons (coding sections of the RNA transcript) are attached together to create the mature mRNA (Nature Education, 2014). Recently there has been an evolving field of research suggesting that noncoding RNA have a crucial role in epigenetics (the field that studies inherited changes in gene expression that do not affect the sequence of DNA) (Wei et. al., 2017). They can also regulate gene expression on different scales (at the genetic level and at the chromosomal level) to exert control on cell differentiation (the process of a cell undergoing physiological adjustment to become another unique cell) (Wei et. al., 2017). For example, various studies have found that long noncoding RNAs (lncRNA) have a crucial role in X chromosome inactivation (Wei et. al., 2017). X-chromosome inactivation occurs due to the fact that, in comparison to males, females carry twice as many X-linked genes in their sex chromosomes (Ahn & Lee, 2008). This additional dosage could be fatal, so during development, one X chromosome is randomly muted in each body cell (cells that are not egg cells) (Ahn & Lee, 2008).

An article published in 2015 addresses the fact that ncDNA and ncRNA, specifically micro-noncoding RNA and long-noncoding RNA, have major relations to various types of cancers (Ling et. al., 2015). The researchers found that non-coding DNA transcribed a variety of long non-coding RNA (Ling et. al., 2015). Areas that contain transposons ('jumping genes' that migrate from one location on the genome to another), repeats and pseudogenes ('fake coding genes,' or genes that resemble protein coding genes but are essentially defective), are important for regulatory activities in the human body (Ling et. al., 2015). In 2017, an article published by Nature Genetics discovered that an abundance of repetitions, specifically recurring regulatory non-coding DNA factors, were found in the genome of pancreatic cancer patients (Feigin et. al., 2017). In 2019, another article released by Nature, found that mutations in regions of non-coding DNA may contribute to an increase in Autism Spectrum Disorder (ASD) risk (Zhou et. al., 2019). Mutations in non-coding regions of the human genome have also been found to lead to Mendelian diseases. In 2020, the article Unusual nature of long non-coding RNAs coding for "unusual peptides", identified that although long non-coding RNA, often transcribed from non-coding DNA, were once thought to be 'junk' they are now known to perform various crucial duties that are essential to cell functioning (Bagchi, 2020). They specifically focused on the recent finding that long non-coding RNA can code for uniquely built peptides (short chains of amino acids) (Bagchi, 2020). This is particularly significant due to the knowledge that peptides are linked to different diseases (University of Queensland, 2017). Peptides or polypeptides are, currently, estimated to be very efficient as drugs for various diseases (University of Queensland, 2017). Researchers have also associated peptide drugs with having relatively fewer side-effects in comparison to drugs derived from other sources (University of Queensland, 2017).

This timeline is meant to expand on how many important discoveries scientists have made once they started asking the right questions and conducting research. It is supposed to demonstrate two things; (1) the pitfalls of not asking questions or challenging ruling ideas and (2) when humans start making inquiries the speed of research output is impressively rapid. However, one of the many things to learn from all this research is that it is quite recent. This is because scientists stalled asking questions or challenging this idea of 'junk DNA' for several decades (Ling et. al., 2015). And although, in just a few years scientists around the world have made fantastic progress, as a research topic this field still has a long road to walk.

If a process as significant as X-chromosome inactivation is mediated by long noncoding RNAs, how 'junk' could noncoding RNA really be (Wei et. al., 2017)? Since a large proportion of noncoding RNA is transcribed by noncoding DNA others may question, is it possible for such a large part of the genome to be functional (Ling et. al., 2015)? What if it is not? We are still questioning the individual function and the extent of functionality of noncoding DNA (Ling et. al., 2015). Not to mention, the characterisations and the mechanisms of noncoding DNA. Some biologists may even question the validity of the ENCODE results (Ball, 2013).

The Evolutionary Perspective: Implications of the DNA theory
How much do evolutionary biologists know about evolution and the phylogenetic tree at the molecular level? The truth of the matter is, not much at all (Ball, 2013). Evolutionary Biology is a broad field of study that explains the

change of generations over long periods of time (Springer Nature Limited, 2021). For generations, these biologists have tried to add molecular evidence to support their well-built theories (Ball, 2013). However, according to the article, Celebrate the unknowns, we know very little when it comes to how the process of evolution (specifically natural selection) plays out at the molecular level (Ball, 2013).

The molecular map of the evolutionary theory has never been clear, especially not now. This fact is evident in many of the conflicting ideas that are recently being published in research journals. In 2017, the article An Upper Limit on the Functional Fraction of the Human Genome, stated that no more than 15% of the human genome could be functioning (Graur, 2017). Another evolutionary biology-based article embraces the idea of junk DNA and claims that this DNA may be used by heterogametic organisms to promote endurance against environmental obstacles and the accumulation of genetic variation without phenotypic effect (Diaz-Castillo, 2016). For context, being heterogametic means that an organism that has different sex chromosomes which results in unique gametes (Diaz-Castillo, 2016). For example, in humans the heterogametic population would be males (Diaz-Castillo, 2016). Where some evolutionary biologists are trying to use 'junk' DNA to construct a more comprehensive view of the phylogenetic tree, others are working tirelessly to prove it worthless (Ball, 2013). As one can probably tell, the evolutionary research community is currently not in the best state, in association with molecular evidence (Ball, 2013). Many ideas are distraught, quite often every single idea that researchers had about the human genome in an evolutionary context is thrown into deep question (Ball, 2013). There were evolutionary theorists who completely disregarded the ENCODE findings claiming that junk DNA is inconsequential to biological processes (Doolittle, 2013). They also criticized the scientists (in the ENCODE project) for claiming otherwise (Doolittle, 2013). But there were also those who used this new idea and tried to find evolutionary relations to all this extra data (Ball, 2013).

At the front, evolution seems like such a firm and unrelenting theory. Biology students may believe they know everything from learning the bare basics of such a complicated and large idea. While their beliefs are admirable, as one can infer from this section, this is the furthest idea from the truth. Another reason for the confusion that has been ingrained into evolutionary biological research is the unrefuted fact that molecular biologists have very little idea of how the human body translates ones' genotype (entire genomic sequence) to their phenotype (external traits) (Ball, 2013). The fact of the matter is that without knowing how genomes turn into their outside characteristics, scientists cannot map the evolutionary timeline at the molecular level. Therefore, what do we know about the genotype to phenotype process? What are we still asking?

At the Surface Level: Genotype to Phenotype

High school students often have a mandatory PTC (Phenylthiocarbamide - chemical compound) test experiment because the test is a relatively simple method of observing single gene inheritance (Merritt et al., 2008). During the PTC test, an individual is given a PTC paper which determines whether they have a sensitivity to bitter taste (Merritt et al., 2008). If one is a taster (if both or one of their parents can taste it they can too, this is also known as the dominant trait) then you do have a sensitivity to bitter taste or may have a range of sensitivity (incomplete dominance where one parent can taste and the other cannot) (Merritt et al., 2008). However, if someone is not a taster (if both of their parents cannot taste it then they cannot taste it either a.k.a. the recessive trait) then they do not have the same sensitivity to bitter taste (Merritt et al., 2008). It took scientists quite a while to discover the underlying mechanism to even this 'simple' single gene inherited trait (Kim & Drayna, 2005). Another example, individuals with cystic fibrosis also carry the recessive allele and it is coded by a gene (CFTR), but did you know that the same disease can be a product of various other mutations (with different degrees of exertion over the phenotype) (Genetic Alliance, 2008)? There are very few genes that code for one specific trait, however, even with these traits scientists do not know or understand a lot of the underlying mechanisms. So, what about the traits that have multiple genes working behind the scenes?

When the Human Genome Project was first released, there were a variety of promising outcomes produced (Crouch & Bodmer, 2020). One of these outcomes was Genome Wide Association Studies (GWAS), these studies use samples of thousands of people to analyze across various genomes for genetic markers that could be associated with different traits or diseases (Crouch & Bodmer, 2020). When these studies were first employed, a promise of associating particular genetic variants to complex traits (coded by various genes) were made (Crouch & Bodmer, 2020). This promise has yet to be kept as most of the important associations had non-essential effects on the phenotypic output and thus finding the gene variants' individual control was not plausible (Crouch & Bodmer, 2020). It became evident that GWAS studies could not be used to fulfil the promise of finding all genetic variants for complex traits, but they were used to develop Polygenic Risk Scores (Crouch & Bodmer, 2020). Polygenic Risk scores are estimations that calculate the genetically natural tendency for an individual to have a specific trait (Crouch & Bodmer, 2020). This helps predict the risk level for a particular group of people to have a trait or, more importantly, a disease (Crouch & Bodmer, 2020). However, using this in a medical context is not yet confirmed (Crouch & Bodmer, 2020). It will take a substantial amount of evidence and research before it can leave the pilot phase (Crouch & Bodmer, 2020). The researcher suggested using individualized studies that focus on rare variants to observe a larger effect on the expression (Crouch & Bodmer, 2020). Polygenic traits however are commonly talked about and thus are much less entairning to explore when compared to pleiotropic traits.

A pleiotropic gene is a gene that affects two or more phenotypic expressions (Masotti et. al., 2019). Pleiotropy is commonly associated with various types of human diseases (Masotti et. al., 2019). In patients with both Holt-Oram syndrome and Nijmegen breakage syndrome, mutations in the specific individual genes that code for important cellular function leads to a patient exuding a collection of symptoms (Lobo, 2008). Although scientists know that various diseases may be a result of pleiotropy, it is extremely difficult to determine the best study method for detecting relevant phenotypic associations with pleiotropic traits (Masotti et. al., 2019). For a while, single gene association tests have been used on GWAS results to detect associations between genes and traits (Masotti et. al., 2019). However, researchers are working tirelessly to develop more efficient methods to study these correlations (Masotti et. al., 2019). One of the methods developed described in the article, Pleiotropy Informed Adaptive Association test of Multiple Traits Using Genome-Wide Association Study Summary Data, uses multi-trait association tests on GWAS results rather than individualized tests (Masotti et. al., 2019). This method has great promise, if or when applied in medical settings in association with disease treatment (Masotti et. al., 2019). For now, however, the general biological research community continues to question the best methods for associating the single gene with the multiple phenotypic variants (Masotti et. al., 2019). They also question how the further development in identification of better data will apply to the treatment of these pleiotropic gene associated diseases (Masotti et. al., 2019).

Summary

At times humans may believe they know enough and stop the progress because there is nothing interesting or applicable to be discovered. However, falling into this fallacy of comfort may be detrimental to a large portion of the population. Protein-coding DNA was discovered and revered for numerous years, with time this progress slowed and gradually came to an almost standstill (Wood et. al., 2019). This eventually led to 20% of proteins being uncharacterized (Wood et. al., 2019). The characterizations of these proteins are essential to many environmental related disorders and should be studied for human development (Wood et. al., 2019). There is also that ruling theories should be questioned for validity because they might not be true after all (Ling et. al., 2015) For decades, scientists believed that 98-99% of the human genome (noncoding or 'junk' DNA) served no function (Ling et. al., 2015). After findings of the ENCODE project, which stated that 80% of this junk DNA served a purpose, this theory was brought into increased scrutiny (Dunham et. al., 2012). There is a continuous stream of research output in recent times, associating various functions, such as regulation of gene expression and X-chromosome inactivation, to noncoding RNA which is mostly transcribed by noncoding DNA (Ling et. al., 2015; Ahn & Lee, 2008). This has important implications for the development of treatment and mechanisms of various diseases (Ling et. al., 2015). However, there are a variety of questions and uncertainty surrounding the debate of the extent of functionality for noncoding DNA (Ball, 2013). Many evolutionary biologists claim that this DNA is irrelevant noise, even stating that it is impossible for the genome to hold more than 15% functionality (Graur, 2017). The question of functionality, mechanism and evolution of noncoding DNA is, nonetheless, very prominent in the Molecular Biology research community (Ball, 2013). A large portion of confusion stems from the mechanism behind genotype being translated to phenotype (Ball, 2013). Biologists have limited knowledge in how most genetic variants work on the environment to construct the phenotype (Ball, 2013). There are very few single gene inheritances in the natural environment, however, even these are under research because of the uncertainty of the underlying mechanism of inheritance (Kim & Drayna, 2005). Polygenic traits make up a large portion of traits, i.e., skin and hair, however even with Genome Wide Association Studies, it is very difficult to associate exact percentage of control over phenotypic trait to most genetic variants (Crouch & Bodmer, 2020). Many researchers are pondering and developing new analysis techniques and study tools to collect more relevant associations (Crouch & Bodmer, 2020). Finally, a wide selection of human diseases stem from pleiotropic traits (Masotti et. al., 2019). Scientists are trying to develop the best method of mapping the associations thereby creating more appropriate treatment methods (Masotti et. al., 2019). There is truly no end to the answers for the inquiry: 'What questions are we still asking about DNA?' The best resolution is to keep asking new questions and consider developing new theories to answer these queries.

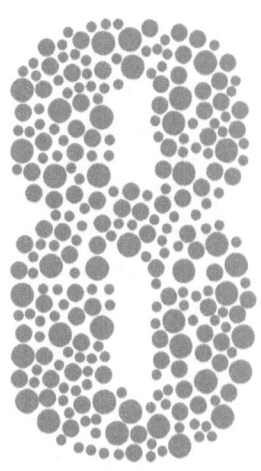

What Ethical Concerns Arise from the Study of DNA?
NEHA SAROYA

Since DNA was first isolated in 1869 by Friedrich Miescher, a Swiss researcher, it began a slowball of advances in the field that have been covered extensively in previous chapters (Pray, 2008). Although there are numerous benefits and scientific advances that have come about due to the study of DNA, it is almost important to take any potential ethical concerns into consideration. It just so happens that the field of DNA research and its implications deal with a variety of ethical concerns, all of which couldn't possibly be covered in a single chapter. However, there are some prominent concerns that should be addressed and known by anyone interested in DNA research. The first of these involves forensic DNA and how DNA evidence that is often used in court to incriminate suspects is not as accurate as many believe it to be due to human error. The second involves germline gene therapy, which constitutes the altering of an embryo's genome before it is born. Despite this gene therapy's potential for advancing scientific research and preventing life-threatening diseases, the ethical concerns surrounding it are so great that human clinical trials in the field are banned in numerous countries.

Forensic DNA
The field of forensic science has been one of great interest in the past few decades due to its ability to provide supposedly conclusive evidence that helps put criminals behind bars. The first case in which this occurred was when Dawn Ashworth, a 15 year old girl from England, was raped and murdered in July of 1986 ("Thirty years of DNA forensics: How DNA has revolutionized criminal investigations", 2021). At the time, a genetics professor by the name of Alec Jeffreys had recently discovered that individuals' DNA patterns could be used

to determine if separate DNA samples were from the same person. As such, the English police recruited Jeffreys to help them find Dawn Ashworth's murderer. Although the police had a suspect that had confessed to Dawn's murder, they weren't sure if he was guilty and wanted to confirm with DNA evidence. It turned out that the DNA from the crime scene was not a match to the suspect, and so the search continued. The police obtained blood and saliva samples from over 4000 men within the age group they were searching for that lived in the area where the murder took place, but there was no luck in finding a matching DNA sample. However, one day a police officer overheard a man saying that he was paid by someone else to provide false DNA samples to the police and upon further investigation the police found their new suspect, Colin Pitchfork. Once they obtained his true DNA sample, Jeffrey's quickly realized that it was indeed a match and Pitchfork was sentenced to life in prison in January of 1988 ("Thirty years of DNA forensics: How DNA has revolutionized criminal investigations", 2021).

Since this revolutionary discovery, DNA evidence has been widely used when available to tie suspects to crimes, or to clear their names. When police have a suspect in custody and have DNA from the crime scene, they will typically request a DNA sample from them ("Advancing justice through DNA technology: using DNA to solve crimes" 2021). DNA cannot be taken from any suspect without consent under the Canadian Charter of Rights and Freedom, unless the police have obtained a warrant (" Steps to Justice", 2021). However, often suspects feel pressured into giving DNA samples, even when there is no warrant, as they believe it will look suspicious if they refuse (" Steps to Justice", 2021). Alternatively, if police have DNA from the crime scene but do not have any suspects yet, they will often run a search through an offender database, which contains DNA sample results from previous known offenders within their area ("Advancing justice through DNA technology: using DNA to solve crimes" 2021). If they are able to find a match, they will know exactly whose DNA was left at the crime scene. This may sound like the perfect way to conclusively tie suspects to crimes and ensure their incrimination, yet there is a lot more subjectivity in this process than there seems at first glance.

After multiple cases of DNA misinterpretation in criminal cases were revealed in North America, Itiel E. Dror and Greg Hampikian decided to conduct a study on "Subjectivity and bias in forensic DNA mixture interpretation" in 2011 (Dror & Hampikian, 2011). Given the high emphasis placed on DNA evidence in the criminal justice system, it is imperative that any possibility of bias must be thoroughly investigated. The study asked 17 independent expert DNA examiners from across North America to interpret DNA from an adjudicated criminal case - meaning that the court had come to a decision regarding the defendant and the case was closed. In the real criminal case, the DNA evidence was a mixture taken from a gang rape case, in which some of the suspects denied any involvement. However, one of the suspects admitted to the rape and was willing to provide testimony stating that the other suspects were also involved, in exchange for a lesser sentence. The problem arose when the DNA evidence is taken into account because according to the laws in the U.S. state where the crime took place, there needed to be corroborating evidence to the admitted rapists testimony in order for the testimony to be taken into consideration. This corroborating evidence came in the form of DNA evidence and those who were analyzing the DNA mixture knew this information. Once the DNA mixture had been analyzed, the examiners concluded that the suspects who denied involvement "could not be excluded". As such, the admitted rapist was able to provide testimony that incriminated the other suspects. When the study was conducted, the 17 DNA experts were not provided with any context and were simply told to analyze the DNA mixture to see if there was conclusive evidence that the suspects who claimed innocence could not be excluded from the DNA mixture. If the original DNA examiners in the real case were analyzing the DNA completely objectively, the independent experts that were not given context should have come to the same conclusions as them. However, this was not the case. In fact, the 17 DNA experts came to differing results even amongst themselves. For one of the suspects in question, only 1 DNA expert concluded that the suspect "could not be excluded", while 4 determined it was "inconclusive", and 12 determined that the suspect could be "excluded". This is in clear contrast to the results of the DNA examiner in the real case who concluded the suspect "could not be excluded". Since all 17 DNA expert examiners analyzed the samples with the same equipment, under the same conditions, and with no prior context, the results of this study bring up alarming ethical concerns. The first being the most obvious, and that is that without context, the majority of the DNA experts disagreed with the original examiner, thus indicating a potential bias in the examiner's methodology. The second concern is that even amongst some of the most experienced DNA experts in the continent, many of them came to different conclusions when provided with a set of evidence. These individual differences demonstrate a level of subjectivity within DNA mixture analysis, that may be related to "training, experience, personality, and motivation". This concerning study calls for further research into potential bias involved in DNA analysis and for DNA examiners to have a higher caliber of training. Thankfully, this study and the mentioned case involve complex DNA and the vast majority of DNA analyses that occur in North America tend to be highly accurate (Leahy, 2021).

However, sometimes even non-complex samples are misinterpreted, leading to paramounting ethical concerns regarding the esteem placed on DNA evidence in the court of law. Multiple cases like this have occured in Houston, Texas, at the Houston Police Department Crime Laboratory. This laboratory analyzed DNA samples for over 500 cases a year and in 2002 it was revealed that "gross incompetence" on part of the technicians led to many misinterpreted samples over the years (Shaer, 2016). This came to light after a tip was given to a news reporter about potential misconduct at said laboratory, at which time DNA samples that had been examined by the lab were sent to independent experts for analysis. Thankfully, when this news was airing, Carol Batie was at home watching on her televsion. She immediately knew that this had to be the reason that her son, Josiah Sutton, was in prison for a rape he did not commit. Four years earlier, when Josiah was just 16, him and his friend were arrested for raping a 41-year-old woman in the back of her car. The victim stated that a few days after the attack, she saw Josiah and his friend walking down the street and told a patrol officer that she had seen her assailants. Both boys maintained their innocence throughout the investigation, and even provided blood samples in an attempt to clear their names. However, this was the beginning of four horrible years for Josiah. Based on the examiner's incorrect interpretation of the DNA evidence, she concluded that the semen found in the back of the victims car matched Josiah's DNA profile based on his blood sample. She also correctly determined that there was no match for his friend's DNA. The following year, in 1999 Josiah was found "guilty of aggravated kidnapping and sexual assault" and was sentenced to prison for 25 years. The reality was that neither of the boys had any involvement in the attack; the real perpetrator was still at large and would not be caught until years later in 2007. Josiah's mother, Carol, always knew in her heart that her son was innocent and following his arrest she even wrote letters to state governors and the Innocence Project to try and free her son, to no avail. Once the crime labratory's routine incompetence had been revealed, Carol reached out to the news station that broke the story and she knew she was finally about to get some answers. Josiah's case was sent to a UC Irvine professor by the name of William Thompson, who took one look at the evidence and immediately knew the examiner had made a grave mistake. The genetic markers from the DNA evidence samples definitely did not match those of Josiah Sutton. After four long years in prison, Josiah was released in 2003. In a shocking ending, the DNA technician who was responsible for misinterpreting numerous samples was fired but then reinstated because her lawyers argued that any mistakes she made were "a product of systemic failures and inadequate supervision" (Shaer, 2016). This further proves the necessity of stricter protocols on who is allowed to become forensic DNA analysts and training that is held to a much higher standard than it was at the Houston Crime Lab.

Germline Gene Therapy

The ethical issues surrounding the study of DNA do not just stop at forensic DNA interpretation. Another hot topic regarding DNA that has been quite controversial is germline gene therapy, which utilizes the CRISPR technique mentioned in previous chapters. Germline gene therapy is used to 'correct' certain genetic variants in parents' sex cells in an embryo so that their child and all future generations are born without those specific genes ("Germ Line Gene Therapy - an overview | ScienceDirect Topics", 2021). Although this could be beneficial for those who are genetically predisposed to diseases that could decrease quality of life and shorten life spans, this issue is much more complicated than it may seem (Anderson, 1985). One of the prominent issues at the time arises from the fact that this therapy is incredibly new; in fact despite it being successfully completed in 2001 (Ferriman, 2001), it is now banned in most countries, including Canada (Kleiderman & Stedman, 2019). As a result, there has been little research done on it in recent years and the long-term side effects of this therapy are completely unknown. It is quite possible that while attempting to fix one genetic mutation, you are unknowingly introducing another genetic mutation into someones' bloodline. Since the sex cells themselves are targeted, any changes made are permanent and would not simply die off with the child of the parents' whose genes were altered. Unless further germline gene therapy was conducted, the altered genes would be passed down to every single future generation. Furthermore, this therapy comes with its own set of risks and has the potential to be harmful to the embryo it is performed on. Besides the obvious risks it may pose to physical health, there are many other factors to take into consideration involving societal expectations.

If germline gene therapy were to become the norm and there were no restrictions as to what parents could alter genes for it becomes a slippery slope that could easily lead to results that look eerily similar to eugenics. Eugenics refers to selective mating in order to purposefully breed for a set of preferable characteristics, in attempts to eventually have societelly deemed 'undesirable characteristics' die off. Given the opportunity, it is likely that many parents would opt to alter their child's genes so that they are born with things that are currently considered disabilities, such as deafness, blindness, physical disabilities, and Down's Syndrome. This would only further divide those with disabilities from those who are able-bodied and would lead to a vast majority of things in society being even more inaccessible for them than they already are ("Is germline gene therapy ethical?", 2021). Despite 22% of Canadians over the age of 15 identifying as having one or more disabilities, our current society is set up in a manner that is only accessible to those without a disability and if this number were to drastically decrease it would make things significantly more difficult for people with disabilities ("Canadian Survey on Disability - Reports A demographic, employment and income profile of Canadians with disabilities aged 15 years and over, 2017 | LDAC-ACTA", 2021). Besides accessibility, creating this divide could also impact how those with disabilities are treated within society. There is already rampant ableism present in every facet of life and by further pushing the narrative that all disabilities are something that must be 'fixed' it could have negative emotional impacts on those with disabilities and could worsen the treatment they receive from others. Furthermore, in countries without universal healthcare or where germline gene therapy would be considered non-essential and not financially covered by the government, this therapy would only be available to the wealthy. This opens up a can of new ethical concerns where the wealthy would have the privilege to decide exactly how they want their kids to turn out. Those without financial privilege would have to continue seeing their children and families suffer from diseases that the wealthy have simply edited out of their genome ("Harvard researchers share views on future, ethics of gene editing", 2021). Given the permanency of germline gene therapy, this is something that would prevail throughout generations. This would eventually cause the wealthy to have an unfair genetic advantage in comparison to the rest of the population.

Summary
Despite all the amazing advances that have resulted from the study of DNA, the basic component that makes up human life, any research conducted on it must be thoroughly examined for the possibility of ethical concerns. When it comes to forensic DNA analysis, mitigating ethical concerns may come in the form of more government regulations on the profession and more training for those who pursue the field. Additionally, bias training could help to eliminate some inherent bias that plays a role in analysis of complex DNA samples. In terms of germline gene therapy, if the study within this field is ever to continue in countries where it is currently banned, there must be tremendous caution taken when deciding what type of genetic alterations should be allowed if made available to the general public.

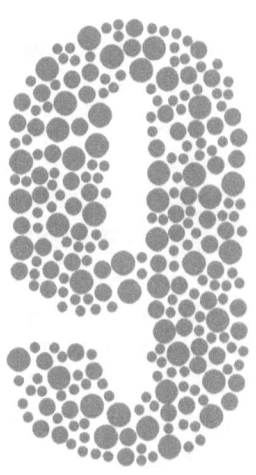

What are Opposing Viewpoints to DNA Theory?
NAZIHAH ALAM

Within history there are many significant findings, which have revolutionized the way future human beings look at a certain issue. But with those findings there also comes critique and concern on the validity, which brings upon a debate. This debate often helps further solidify the knowledge of the concept while exploring the various possibilities. When a significant finding can withstand the critique and concerns, it can become much stronger. One of these significant findings was the DNA theory by James D. Watson and Francis Crick. Their work was revolutionary, but it should also be noted that there were other scientis's work from the past that were helpful in coming to their conclusion, more can be read in chapter one. After many decades of research, the duo were able to configure a double helical structure which to this day is used in all levels of academic literature to explain Deoxyribose Nucleic acid, also known as DNA. It is to be noted that this theory did not come short of opposition when it was first published.

Watson and Crick's model of DNA was highly regarded as it was able to support four main principles that are needed to be genetic material, which were how it explained the process of genetic material replication, the uniqueness of the genetic material, how the DNA strands were able to hold onto information and it also explained how genetic material was able to mutate (Portin, 2014). Their findings allowed them to achieve the Nobel Prize in 1962 for Physiology and Medicine.

Skepticism Versus Denial

It is important to note that there is a difference between skepticism in regards to significant findings versus the denial of significant findings. In the case of skepticism, a scientist yields to find information and reach a conclusion through sufficient amounts of evidence (The Nature of Science. Climate Science Investigations South Florida - The Nature of Science, n.d.). This evidence does not mean that the thing being tested is certain, however it does give evidence for further research and possibility. On the other hand, denial is when a scientist refuses to give up despite there not being scientific evidence to back the data up (The Nature of Science. Climate Science Investigations South Florida - The Nature of Science, n.d.). Skepticism within the scientific community is therefore helpful in solidifying understanding within the subject (The Nature of Science. Climate Science Investigations South Florida - The Nature of Science, n.d.).

Skepticism within the Scientific Community - Max Delbrück

However, even with their evidence there was still skepticism within the scientific community. One of these people would be Max Delbrück, who was German-American biophysicist (Hernandez, 2017). Although he originally worked in the field of physics, he was also very interested in biology which led him to merge these topics. He would also be later awarded the Nobel Prize in 1969 along with Salvador Luria and Alfred Hershey in Physiology or Medicine (The Nobel Prize in Physiology or Medicine 1969, n.d.).

Delbrück agreed with the structure of DNA that was proposed by Crick and Watson, but he did further question how the DNA replication process could occur without getting tangled (Hernandez, 2017). Watson and Crick had developed the semi-coservative method of replication. This is where the double helix structure of DNA would essentially unzip, where the base pairs would be separated. That unzipped strand would become a template for the new DNA to replicate from. The DNA that becomes the template is called the parent, while the new forming DNA is the daughter. Delbrück found this process hard to believe as it seemed as though the DNA would get tangled in the process of replication and cause issues (Hernandez, 2017). He instead created his own hypothesis on what could actually be happening during replication.

Genetic Replication Concerns by Delbrück

In his paper, "On the Replication of Deoxyribonucleic Acid (DNA)", he first explored three possible ways the DNA strands could be unzipped before entering the replication stage (Hernandez, 2017). In his first method, he described the DNA strands would be vertically pulled, essentially sliding past one another in order to unzip. However, this idea once again could lead to tangles. He next further considered Watson and Crick's solution of semi-conservative method, which although was a possibility still did not seem to avoid the tangling issue (Hernandez, 2017). This brought him to his third idea which would later be known as a dispersive replication method. This was arguably a bit more of a complicated process, where the DNA would have specific breaks and reunions done at each twist of the structure (Hernandez, 2017). Essentially the DNA would have a mixture of both parent and daughter strands as at specific points, the DNA would break and would then be used as a template to replicate new DNA, and then it would rejoin with the rest of the DNA (How DNA Replicates, n.d.).

There were scientists during that time that looked at both sides of the debate, in particular Matthew Meselson and Franklin W. Stahl was able to determine that DNA did replicate through the semi-conservative method through their experiment (How DNA Replicates, n.d.). The experiment consisted of growing the bacteria Escherichia coli, also known as E.coli, in different levels of isotope of nitrogen (levels such as heavy, medium and light). At each level, the E.coli would be left to reproduce, which is done quite quickly, in each generation could approximately concur in twenty minutes (E. coli – the biotech bacterium, n.d.). Through the density gradient of centrifugation, Meselson and Stahl were able to determine that the semi-conservative model was indeed the method at which DNA would replicate. The study with E.coli could trascened to other organisms as well (Khan Academy, n.d.). They were able to publish their findings in the Replication of DNA: A History of "The Most Beautiful Experiment in Biology,". The issue with the dispersive model proposed by Delbrück was that over generations of DNA replicating in that method, the strands eventually decrease in density, which would make the DNA more susceptible to mutations (Khan Academy, n.d.).

Side by Side Model
There were further papers that also delved in other possibilities on how the DNA strand was structured. In the 1970s there were also many papers circulating which critiqued whether the DNA strands would stay beside one another without coiling as Watson and Crick mentioned in their proposal (Portin, 2014). Some of those papers claimed that the DNA would be placed in a tape-like fashion which would avoid tangling, which was as mentioned before a large concern (Portin, 2014). This formation would be easier when having all the DNA together. Some of the models that were designed to support the tape-like structure of DNA also seemed to fit better with the x-ray crystallographic information that was available (Portin, 2014). Despite these concerns, Crick and his colleagues were able to debunk the "side by side model" by adding more information on how the coiled DNA's double helix model actually worked.

Concerns on the Double Helix Model Itself
There were also other concerns about the double helix model itself. One of

which was considering whether the two strands of the double helix were actually pushed together, which was also known as plectonemic coiling, rather than wound together. This was once again proved to be false by Crick and Watson later on.

Summary
Having opposition within the scientific community is essential in fully understanding new discoveries. When a significant finding can withstand all the critiques, it is made much stronger, in this case the DNA theory was able to maintain well woven into all levels of education because of all the supporting evidence. Children in high school to Students in their pHD can attest to this model and it opens the gate to learn more about DNA as a whole.

What Misinformation or Conspiracy Theories Exist Regarding DNA?
POOJITHA PAI

With the advent of the internet and high-speed worldwide communication systems, it has never been easier for misinformation and conspiracies to spread to a great degree in a rapid fashion. One of the challenges when tackling misinformation is that it is complex to define in the ever-changing world of scientific discoveries. After all, something that was believed to be true by the medical community at one point can now be classified as misinformation with the most up to date information. For the purposes of this chapter, misinformation is defined as "information that is contrary to the epistemic consensus of the scientific community" (Swire-Thompson & Lazer, 2020). Rampant misinformation can cause many individuals to distrust the scientific community at large and engage in behaviors that can put themselves or others at harm, a phenomenon that was highlighted in the COVID-19 pandemic.

A related concept, a conspiracy theory, is an idea or an explanation of an important event that often includes secret involvement of powerful and malevolent groups. Even in this era of endless information, individuals struggle to find causal explanations for major events or phenomena that occur in their lifetime. Many are thus drawn towards conspiracy theories due to a desire for understanding why or how an event occurs, to achieve a sense of control and security over information, or to maintain a positive self-image (Douglas et al., 2017). These conspiracy theories are not limited to major events and can be directed towards information that have been provided by major organizations. One area that conspiracy theories, especially in recent times are focused on, is one that has fascinated human beings for millennia

– the human body. More specifically, these conspiracy theories question the information that is given out by those they believe to be 'malevolent' groups (e.g. medical community).

DNA, the blueprint of life itself, is not immune to both misinformation and conspiracy theories. In fact, the absence of complete understanding of how these miniscule molecules operate or 'code' such large quantities of information has fuelled many individuals to turn to alternative, and often misinformed, explanations and theories. In this chapter, I will be discussing significant conspiracy theories and large amounts of misinformation that surround DNA and its applications. I will attempt to highlight their origins, consequences, and the reasons why they persist.

Misinformation and Conspiracies Surrounding DNA
Races and Racism
Across the world, humans share about 99.9% of their DNA. The variability shown through skin colour, hair texture, or facial features makes up just 0.1% of the thousands of traits that define us as 'human'. Yet, within this 0.1% of genetic variability, humans throughout the years have sought to classify, categorize, and place in a hierarchical fashion. This 0.1% has been the reason why science has been misused over centuries to craft the race concept that still shadows contemporary genetics.

The first scientific system to classify humans based on race was established French physician Francois Bernier in 1684, and then further elucidated upon by Swedish naturalist Carl Linnaeus in 1758. This system divided all humans into four groups: europaeus, americanus, asiaticus, and afer. It also not only described physical characteristics such as skin colour or hair texture, but also made judgments on each race's characters and personalities. In the 19th century, Charles Darwin introduced the idea of evolution, which theorized that all humans originated from the same ancestor. However, in line with the thinking of his times, he asserted that natural selection would place the white races as the fittest as they were the most evolved. Francis Galton, Darwin's cousin, became an early proponent of the eugenics theory which was the notion of improving the human species by only selectively mating people from suitable races with 'desirable' characteristics. Eugenics since then became the root of many extreme theories that influenced individuals to enact the Nazi policy of racial 'cleansing' (Mohsen, 2020).

Currently, our understanding of genetics and variability in genes is far greater than before. A landmark study by Stanford scientists in 2002 examined diversity of alleles in populations across the world. They found that a majority of alleles were shared among people of different regions. In fact, this study and many others have shown that there is more variability in individuals within a particular race than there are between individuals of different races. In summary, the notion that individuals from different races were inherently distinct from those of another race is simply not supported from a genetic standpoint.

However, even after almost 20 years after this study's publishing, the idea that races are genetically different to each other has thrived. The recent craze surrounding genetic ancestry testing (GAT) organizations such as 23andme are a testament to our sustaining fascination towards the race concept. Ancestry testing can be a powerful tool for those without the luxury of a family history. For example, it can be particularly meaningful for African Americans who hoped to reconnect with their roots to "help heal the genealogical wounds inflicted by the Atlantic slave trade" (Sachs, 2019). However, GAT can also counterintuitively become a tool for exclusion, xenophobia, and racist doctrines.

A recent study showed that white nationalist groups use genetic information through the use of GAT to not only justify but also to fuel their beliefs of superiority over other races. GATs are used to determine their 'purity', which give them opportunities to join white nationalist communities and online platforms. On analysis of their discourse on the online forums, it was found that white nationalists actively engage in genetic, statistical, and anthropological knowledge of human diversity. However, particular pieces of information are cherry-picked in order to fit their own narrative of racial boundaries and hierarchies (Panofsky & Donovan, 2019).

Additionally, they engaged in conspiracy theories to reconcile their lack of 100% ancestry with their 'pure white' self-image. One particularly prevalent conspiracy was that the Jewish people controlled the ancestry test organizations, and would "sprinkle some non-white DNA" to anger and weaken the white nationalists (Panofsky & Donovan, 2019).

There are also people on the other side of the spectrum. In this, people use GATs to identify themselves as ethnically diverse as possible to absolve them of blame for their ancestral racial inequality (Sachs, 2019). It can even be beneficial for politics, as a senior US senator proclaiming to be of Cherokee descent turned out to be between 1/64th or 1/1024th Native American, leading to uproar from the community at large (Mcmaster, 2018).

However, in both cases, individuals have engaged in the false conflation of their genetic make up with some aspect of their cultural identity, personality, or morality - an idea that began with Bernier and Linnaeus. Even after the horrors of the holocaust and the Rwandan massacre, this construct of 'race' permeates throughout society – in science, mass media, politics, etc. – without any scientific precedent.

ONE-GENE, ONE-TRAIT

The one gene, one-trait hypothesis states that one gene codes for one specific characteristic of the organism. While there are some traits such as freckles and dimples that do follow this hypothesis, traits such as eye colour and tongue rolling are clear exceptions to this rule (Cite myths of human genetics). Many organisms also show pleiotropy (when one gene affects multiple traits) and polygenic traits (when multiple genes code for one trait). This was originally a competing hypothesis to the one-gene, one polypeptide hypothesis, both by American geneticist George Beadle. Thirty years prior to him receiving his Nobel prize, he believed that each gene coded for a specific characteristic in the maize plant. However, he later discovered that genes in fact coded for different polypeptides (often enzymes) and accepted the one-gene, one-polypeptide theory, which since then has undergone substantial sophistication.

While the consequences of this misinformation itself is hard to elucidate, it does shed a light on the plight of scientific education. Nonetheless, this overly simplistic model of genetics has been perpetuated in most high school and college-level biology classes, leading many to undermine the complexity of genetic information and the influence of a single gene on a person's phenotype. University of Delaware Genetics Professor John McDonald in his book, Myths of Human Genetics, commented, "It is an embarrassment to the field of biology education that textbooks and lab manuals continue to perpetuate these myths". This misinformation causes people to believe in a fundamentally flawed principle about the nature of the relationship between DNA, the proteins it codes for, and the resulting trait seen in an individual.

This may be one of the reasons that many pop-cultural references to DNA and genes are so simplistic. In how many movies or TV shows have you heard medical characters look at the results from genetic tests and say, "Thank God! You don't have the gene for this disease!"

In reality, this often is not the case. Firstly, genes here are a misnomer. Most diseases are not caused by the presence or absence of a certain gene. For some diseases, such as Huntington's Disease or Duchenne Muscular Dystrophy, it is the presence or absence of a mutation (a mistake) in a gene – a gene that all humans have – that determines if the person will have the disease. Secondly, most common diseases such as type-2 diabetes (Udler, 2019), cancer (Tsaousis et al., 2019), and asthma (Bijanzadeh et al., 2011), require a complex interplay of multiple genes and multiple mutations that cause the diseases to be presented.

Genetic Modification by Clustered Regularly Interspaced Short Palindromic Repeats (CRISPR)

As explained in previous chapters, CRISPR is a tool that allows scientists to precisely insert or excise specific parts of the DNA into an organism (Hsu et al., 2014). With this tool, scientists are able to permanently modify genes of an organism's living cells, which can potentially be carried over to the next generations of that organism. To an average person, CRISPR seems like a technology from a science fiction movie. The common saying by skeptics and critics of CRISPR is "We shouldn't play god!". Are there fears unfounded? After all, this is the first time that mankind has been able to play around with genetic code, right?

Wrong.

Humans have been tampering with the genes since the beginning of civilization thousands of years ago. We have identified desirable traits in plants and animals around us and selectively bred them such that only those traits would be passed onto the next generation. It is artificial selection that turned bitter and tiny teosinte to the domesticated corn we enjoy today, that turned the wild auroch into the excess milk-producing dairy cow, and that turned majestic and wild wolves into chihuahuas.

Genetic modification is nothing new in our history. But never before have we been able to engage with genetics with such precision and over such little time. Before, it would have taken generations to get a plant or animal with all the desirable traits, but with CRISPR this process can be immensely sped up. However, with the complex nature of gene editing and the processes mainly being understood by only those in the scientific community, an average person's contact with CRISPR is often through sensationalized news articles propagated through social media and mainstream channels. It is no surprise that there are a whole range of myths and misinformation that have been circulating around CRISPR.

1. Creating designer babies

As much as science fiction would like us to believe, we are nowhere near being able to create genetic superbabies. As highlighted in the previous section about the fallacy of the one-gene, one-trait hypothesis, it is incredibly complex to determine a gene that is associated with each trait, not to mention that there are factors other than genetics that can play a role in presenting one's traits. Factors such as nutrition, environment, socio-economic status can have a complex interplay with genetics to determine one's intelligence, athleticism, or talent in a particular field.

2. CRISPR can cure all genetic diseases.

The most common application that is talked about for CRISPR is the removal of certain genes that could cause diseases. We say news articles that say something along the lines of, "CRISPR will one day be able to cure all genetic disease". This is not the case. There are about 10,000 genetic diseases that are caused by single gene mutations – like Huntington's disease or Duchenne Muscular dystrophy. However, thousands of genetic diseases have multiple genetic factors – not all of which are inherited. Cancer, for example, may be caused by inherited mutations (Hodgson, 2008) but can also be the result of "de novo" or new mutations that may have been accrued over a lifetime in response to the environment (Acuna-Hidalgo et al., 2016). Since we cannot isolate the specific genes that cause this disease, it is impossible to use CRISPR to excise those parts of the genome.

Genetically Modified (GM) Foods

As mentioned before, genetically altered foods have been a part of our history since the beginning of civilization. Selectively breeding desirable characteristics have caused plants and animals to change their traits significantly. Modern techniques allow for this process to occur at a much more rapid state. Discourse surrounding GM foods has been largely negative in the media leading to a large amount of misinformation and conspiracies. As such, this area of genetic modification warrants its own section to help clear some common misconceptions surrounding it.

1. GM foods are harmful to humans and the environment

Common concerns about GMOs include adverse health effects due to toxicity, allergies responses, and differences in nutritional content. However, GMOs as we know it have been a part of the agricultural and food industry for the past 20 years. In this time, they have had a safe track record. The American Medical Association and the World Health Organization have both conducted independent research and have concluded that GM foods are safe for consumers. Similar to how genes from the non-GM foods do not cause us to turn part-plant, new genes introduced to GM crops for longer shelf-life, increased nutritional content, pest-resistance, etc. will not be able to harm the consumer (Norris, 2015).

In fact, not relying on GM foods can force farmers to use more toxic pesticides to preserve the food – exposing more people to harmful chemicals, increasing production costs, and limiting accessibility of the food (Oliver, 2014).

2. GM foods are unnatural

In addition to understanding that artificial selection is something that we have engaged in for millennia, it is also important to understand the origins of the mechanism for genetic modification. Artificial genetic modification is based on very natural processes done by bacteria and viruses to transport genes between species. With modern tools, we are able to specify the target gene and do this at a rapid rate.

Additionally, any new genes added to plants are copies of pre-existing genes that exist in nature. Scientists are unable to construct completely new genetic sequences in order to achieve the traits that they desire. Therefore, even if new genes are introduced, they are taken from existing organisms – making GM foods natural.

3. Organic foods are better than GM foods

Unfortunately, this claim falls short as they seem to be comparing what is analogous to apples and oranges. GM foods are a method of breeding plants for food production. Organics, on the other hand, is a method of crop cultivation. Therefore, they are not mutually exclusive. In fact, some GM foods could potentially be safer than organics because some growers of the latter are allowed to use harmful toxic chemicals pesticides, while growers of pest-resistant GM foods do not have to rely on them (Husaini & Sohail, 2018).

4. The 'fish-mato' myth

The ongoing rivalry between organic and GM industries have also led to a lot of fear-mongering against GM foods (Prakash, 2015). One common piece of misinformation is the myth of the "fish-mato". This originated from the fact that in the 1990s, there were some experiments that aimed to introduce a fish gene into a tomato in order to make the tomatoes frost tolerant.

'This tomato will taste fishy.'

'I have a fish allergy and I cannot eat this!'

These were some of the many concerns asked by the public due to the fear caused by misinformation. In reality, this tomato was never put on the market but gave many skeptics ammunition against GM foods and is a myth that continues to persist to this day. These concerns were caused by a fundamental misunderstanding of genetic modification and genes. Like was described by the one-gene, one-trait hypothesis, there are some characteristics that are controlled by only one gene. This is true of the gene from the Arctic flounder that lives in the ocean and codes for a protein that prevents its blood from frosting. However, splicing that gene into a tomato (which was unsuccessful), would even theoretically only give tomatoes that specific property. It would not give the tomatoes any other properties – such as the proteins that would give it a fishy smell, taste or would cause allergies for people with fish allergies (Schwarcz, 2017).

5. GM FOODS BY THE CORPORATE WORLD TO CONTROL DEVELOPING NATIONS AND FOOD SUPPLY FOR THE WORLD

GM foods have often been largely produced by those in 'developed' nations. As a result, many individuals think that making genetically modified foods would allow corporations to control 'developing' countries' food supply (Utinans & Ancane, 2014; Veltri & Suerdem, 2013). Fortunately, this is not the case. Developing nations have increasingly been using GM organisms to provide crops to their citizens, with them producing 56% of GM crops in 2013 (Oliver, 2014). In the developed countries, consumers often do not feel the difference in prices and quality of crops with GM food, as previous methods of growing crops yielded similar results, albeit with greater damage for the environment (Herrera-Estrella & Alvarez-Morales, 2001). However, in developing nations, usage of GM crops has actually helped farmers grow crops for a lower price as they were no longer paying for a vast number of chemicals such as pesticides and herbicides.

COVID-19 Vaccinations

The last section of this chapter is perhaps the most topical, as it has been written during the COVID-19 pandemic. Vaccine hesitancy is at an all time high as the hidden pandemic, misinformation, has run rampant during this difficult time. Skeptics of the vaccines have had many claims about the lack of safety, the speed at which this vaccine was produced, and the long-lasting effects of vaccination on the human DNA.

1. THE VACCINE WAS MADE TOO QUICKLY, SO IT MUST NOT BE TESTED PROPERLY

These RNA-based vaccines have been around for decades, being researched and developed for diseases such as influenza, Zika, and HIV. Years of previous research on RNA vaccines as well as similar previous viruses such as the Severe acute respiratory syndrome (SARS) and Middle East respiratory syndrome (MERS) viruses allowed scientists to rapidly start producing vaccines

(Ball, 2020). Additionally, the removal of the red tape and bureaucracy involved the vaccine production aided this process as well, not the cutting of corners. We have seen that these vaccines have been through all the trials and phases that are required for a vaccine to be approved for the public (Levine, 2020).

2. The vaccine will alter a person's DNA

One particularly common claim is that the vaccine will change your DNA permanently. As explained in chapter 4, the COVID-19 vaccine, like many other vaccines, contain a small part of the organism's RNA or protein. These fragments, when injected into the body, cause individuals to have a usually mild immune response. At the end, the body naturally produces proteins called antibodies. These antibodies would then have the ability to protect the body against the virus if exposed to it in its native state. This process is completely natural as our body reacts the same way with any exposure to a new virus. When exposed, the immune response results in antibodies and cells that are able to produce more antibodies in the future. As a result, when exposed to the virus again, these antibodies are ready to be deployed to destroy the infection (Pulendran & Ahmed, 2011). At no point in this process is the normal human DNA altered.

Summary

Misinformation and conspiracy theories have become commonplace in our society and has the ability to spread rapidly owing to the internet. In this chapter, I have identified numerous myths surrounding DNA and its related components – from the construct of race to genetic superbabies. The dangerous effects of this misinformation have manifested through racism, vaccine hesitancy, and distrust of the scientific community as a whole. Clear communication of scientific information from trusted professionals, improving populations digital literacy, and holding large corporations (such as Facebook) that allow such discourse to take place on their forums accountable are some of the steps that must be taken in the coming years. The tenacity and widespread acceptance of these theories are frightening and should serve as a wake-up call for the scientific community. Unless we want to lose the credibility and trust of our population, we must step up to the challenge of tackling misinformation with equal tenacity and perseverance.

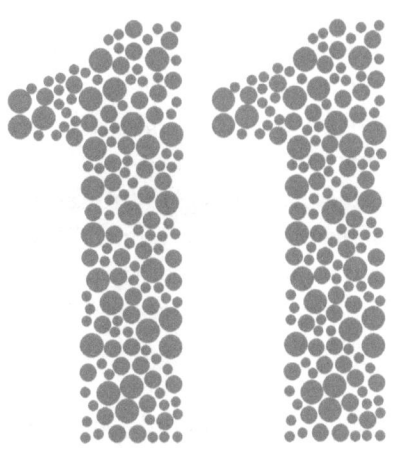

How is DNA Talked About Commonly in Popular Culture?
HAYLEY ZHONG

Deoxyribonucleic acid, or more commonly known as DNA, is a topic that many of us may not have learned about in depth, if at all. Despite this, it is a topic almost all of us have some sort of exposure to. Its prevalence in mass media and pop culture has infused bits of knowledge about DNA into each of our repertoires, each containing our own interpretations of information presented to us about DNA, whether that information be direct or implied, or come from a class, a film, a show, a book, or the news. It is a popular theme explored by writers and often used to expand the breadth of what is familiar into unchartered territories, steering the narrative towards what may not be known to us present day.

In the 2021 episode of the popular reality TV show Keeping up with the Kardashians, which follows the lives of the Kardashian-Jenner family, the three Kardashian sisters, Kim, Khloe, and Kourtney compete in a Spartan obstacle course with their Jenner half-sisters, Kendall and Kylie (Seacrest et al., 2021). The family resorts to this obstacle course challenge to settle Kendall Jenner's claims of blood tests indicating to her that the Jenner family DNA shows that they are genetically gifted in athletics, hence, athletically superior to their Kardashian half-sisters. For context, Kendall and Kylie Jenner are the biological daughters of Caitlyn Jenner (formerly known as Bruce Jenner,) the 1976 Olympic decathlon gold medalist. In Caitlyn's Olympic career, she set three consecutive world records, earning herself the unofficial title of the "World's Greatest Athlete" (Sawyer, 2015). Following the challenge, despite only Khloe, Kourtney and Kendall competing due to the others not feeling well, Kendall does in fact complete the Spartan obstacle course in the fastest time of the siblings that competed.

The vague portrayal of DNA in Keeping up with the Kardashians raises a host of concerns about the delivery of information about genetics in mass media. For starters, there has never been a gene discovered linked to general athletic ability (Brym, 2019). From forensic analysis in crime shows to designer babies like dystopian futures, DNA has made its appearance in its fair share of mass media, including TV series, films, and novels. Despite academic concerns about its accuracy as depicted in popular culture, stories built around the effects or changes of DNA are continuing to grow and those beliefs are infiltrating the lives of individuals, being applied to their day to day lives.

According to Oxford Languages, popular culture is "culture based on the tastes of ordinary people rather than an educated elite" (Simpson, 1997). Film, radio, television, books, slang, fashion and sports are all elements of popular culture. Media is a key transmitter of popular culture. With the average Canadian spending over 69 percent of their time awake interacting with sources of people such as the Internet, television, radio, and newspaper, its ability to influence popular culture and people should not be overlooked (Young, 2016: 5, 8). Humans are not empty vessels for the media to flow ideas into without resistance. We have the conscience and freedom to filter the messages and select which we internalize. However, messages that are most widely spread and appealing like blockbuster hit movies, tend to be the ones people choose to embrace (Brym, 2019). Pop culture has the ability to bring people together over commonalities and actually shapes how we understand reality. In regards to DNA and genetics, the problem comes with false ideas being perpetuated. When we see DNA being portrayed in a certain way over and over again, being applied in a multitude of scenarios and mediums, that specific portrayal begins to resonate as fact.

In the 1993 movie adaption of Jurassic Park, scientists extracted dinosaur DNA and filled in the miss genomes with frog DNA. As a result, they cloned a dinosaur that was physically and behaviourally identical to its ancestors, while having the same reproductive practices as frogs (Spielberg, 1993). In this context, gene expression is extremely simplified and takes credit for traits like behaviour that have an aspect of nurture. These movies are widely distributed and more likely to be seen by the general population than a textbook on genetics may be because of their entertainment value. This misleads many to accept the false conclusion that the relationship between genes and the characteristics they express is rigid, linear and lacking variability – that the presence of a particular gene will always produce a specific characteristic.

Genetic technology has been incorporated in horror, science fiction, and dystopian plots for years, like in Frankenstein, where Viktor Frankenstein creates life in the form of a human-like monster or Brave New World, where the government manufactures different groups of human castes to fulfill different roles in society. Movies, television shows, and books have been criticized by scientists for eliciting irrational fears of genetic technology with tools such as exaggeration and themes of anti-science (Vackimes, 2010). This fear causes people to interpret the "science" in the story in a shallow and reactionary manner, and reinforces a doubt or fear of science itself (Haran & O'Riordan, 2017). This fear makes its appearance in everyday conversation and over the news. An example of this includes dramatic imagery such as "Frankenstein foods" when discussing genetically modified foods, "Frankenstein" being used in a negative connotation (Huges & Kitzinger, 2008).

With mass media and popular culture playing such an integral role in shaping the public's perception, it is important to recognize its effects and role in shaping the public's perception and opinions on topics, including that of DNA. The spread of misinformation is frequent due to the great quantities of the use of DNA infused with creative liberties in popular culture. People tend to neglect the complexity of genes, believing that they are the sole determinant of one's future like in Jurassic Park, and struggling to separate fact from fiction.

These forms of media have grown in popularity because of the preference for drama over scientific reality. Because of how DNA is frequently depicted, many believe in the oversimplified model of DNA that a specific gene will without a doubt code for one particular trait, no matter the state of the other factors involved. Along with this, for entertainment purposes, DNA is wrongly used to fully explain qualities such as personality and intelligence, arguing that the right combination of genes will lead to specific attributes in personality or degrees of intelligence. Without a better understanding of science, coupled with the same repeated narrative being presented, people have a difficult time recognizing when dramatic license is being taken so it is important to see the media's portrayal of DNA for what it is – entertainment.

Within the realm of sports, it is not uncommon to hear phrases such as "black people are better at sports than white people are." This implies that there is a biological difference in the DNA of black people that codes them to be superior athletically. Seeing that African Americans make up two-thirds of both the National Football League and National Basketball Association, the numbers seem to support this statement (Entine, 2000; Lapchick, 2017). However, when other factors are introduced, like the fact that no gene has ever been identified to be linked to general athletic superiority and how black athletes do not dominate in many other sports like hockey, cycling and gymnastics, that statement begins to look less like fact (Brym, 2019). With popular culture, it is widespread because it is easy to understand and it is far easier to make claims about biological reasonings because of our belief in our oversimplified understanding of genetics, rather than explain the plethora of environmental and social factors that nurture this success like how those faced with widespread discimination turn to professional sports due to other means of upward mobility (Brym, 2019).

Overall, the growth of the prominence of false information about genetics in popular culture has added to the pool of people who believe in the power of nature over nurture. Over time, for the average person, DNA has become more than just the basis of a plot and analyzing DNA has become more accessible to individuals through home tests like 23 and Me. The 23 and Me test requires participants to collect a sample of their saliva in a tube that is then sent back to their labs in California to be tested. The test claims to help connect one's ancestry and identify their "true" race. In one of their ads, a woman who once identified as Hispanic began to select "other" on forms that asked about her ethnicity because her 23 and Me test labelled her as 33 percent Indigenous and 31 percent Iberian. It advertises this as the true meaning of the percentage breakdown despite it actually representing the probabilities of where one is likely to have relatives from based on samples 23 and Me agents have collected from regions around the world (Molla, 2019).

The idea that this test can determine race, and break down the make up of one's ethnicity is problematic because they convey the message and encourage a way of thinking that races biologically have clear differences between them. This mindset falls in line with views of biological determinism. Biological determinism is the belief that we do not actually have free will to choose who we will be and that our DNA essentially determines our destiny (Resnik, 2006). Biological determinism undermines sociological factors by upholding the idea that different races are different by nature, hence perpetuating beliefs of a hierarchy of races, supporting racists and supports of eugenics (Hein, Dar-Nimrod et al., 2017).

This test attempts to look at patterns within races and can have people fall into the trap of believing that correlation equals causation. To try to look for patterns within races, 23 and Me sends agents to regions to buy saliva samples from people who claim to be part of a specific ethnic group. This process is inherently flawed because of the low likelihood of someone being purely one race due to thousands of years of intermarriage and internal migration. In some areas, members of the reference sample may have also lied about their race to earn the fee for their saliva, making the likelihood of a homogenous sample group being very low. For example, researchers discovered that black people in North America are more likely to have cancer than white people. After hearing this, the general assumption is that due to biological reasons, there is some genetic makeup in black people to make them more likely to have cancer, or in white people to make them less likely to have cancer. However, the full picture is not painted to be able to draw a clear conclusion. Black residents are more likely than white residents to live and work near environments that produce byproduct pollutants that cause genetic mutations which lead to cancer (Brym, 2018). This demonstrates that it is simply incorrect to trust DNA entirely because of the complexity of one's environment and the environment's ability to impact an individual. The popularization of these home DNA tests have come along with an increase of disputes over genetic determinism (Condit, 1998). Its oversimplification in the mass media in advertisements, and the product itself undermines the idea of free will, neglecting the context of one's environment on their development in favour of deterministic ideologies.

More often than not, popular culture makes DNA far more simple and gives it far more influence in determining complex traits like personality and intelligence than it may actually have. It has been used in such context so frequently that many have accepted it as fact, viewing genetic engineering as almost magic. The idea of the rigid structure of DNA and one genome directly influencing one particular trait is one that many people accept as reality, so much so they may not believe that humans share about 99 percent of our DNA with chimpanzees (Gibbons, 2017). Nevertheless, mass media's contributions towards popular culture in the discussion of DNA is not always detrimental. When used to spark conversation, interest and dialogue with the public, it can be quite effective in creating 'teachable moments.' Like how Gattaca encourages ethical questions regarding genetic discrimination or how Spiderman could inspire someone to learn about genetic mutations. Pop culture's depictions have value and at the end of the day, perspective is what shapes our takeaways. Without perspective, applying the ideas brought forth about DNA by popular culture could be like opening Pandora's Box, like mentioned when addressing possible implications of the 23 and Me test fueling racism and genetic determinism.

Overall, pop culture is so widespread that it has the effect of bringing people together. It likely has more influence than formal science education to shape most people's understanding of science because a Facebook post likely connects with more people than a scientific textbook. Being able to teach others to not accept fiction as fact will be far more effective than teaching extensively about DNA to those who may not necessarily need such depth of knowledge. At the end of the day, droning on about the inaccuracies in popular culture does not really accomplish anything and a better solution is to embrace the media's depiction of DNA for what it is – entertainment.

Where is DNA Theory Headed in the Future?
Juwairia Razvi

The first discovery of genetic material, which would become known as DNA was by a Swiss physician named Friedrich Miescher in 1869. He was interested in learning about the material composed in the cell nucleus. He collected pus from a nearby hospital and purified it from the presence of leukocytes. What was left was a pure organic substance which he named nuclien. He found that nuclein contained no sulfur and high amounts of phosphorus. In the years to come, scientists made many discoveries of the molecular content of DNA. In 1893, Alberecht Kossel noticed that nuclein was a part of chromatin, the substance which composes chromosomes together with proteins. However, the molecular structure of the DNA has been a subject of debate for many decades, and many theories of the protein component being disproven scientific research. In 1953, Watson and Crick discovered the molecular structure of DNA as a double-helix model and published the DNA theory, explaining the genetic implications of the model. This model explained the four properties of genetic material: replication which is essential for reproduction of life, how specificity is preserved in the duplication process, how the DNA stores information and the ability for genes to mutate (Portin 2014).Watson and Crick were awarded the nobel peace prize for their crucial discoveries in DNA theory. This extremely precise scientific discovery has paved the way for innovative and significant scientific developments such as genetics, agriculture, and medicine.

Use in recent history and modern times

Throughout the 1970s, there was significant advancement in gene-editing technology. Bacterial restriction enzymes were discovered, and their ability to cleave DNA at specific sites made it possible to isolate, clone, sequence and transfer genes artificially made possible many technological advancements (Portin, 2014). Through the recombinant DNA method, DNA fragments containing genes of interest could be extracted from a donor organism, joined with vector DNA (usually bacteria plasmid) and inserted into another organism. Recombinant DNA has been used in discovering insulin treatment for diabetic patients as well as GMO and the commercial food industries (SLU, 2016). Starting from the 1960s, the technology has allowed scientists to select for certain genes in plants, transfer genes from other organisms, and attain high yield and high quality of plants. The recombinant DNA method also allowed scientists to transfer the human insulin gene into bacteria in order to produce insulin. This could be harvested and purified to produce insulin medications for patients with Type I diabetes. The widespread availability and use of this treatment has allowed individuals with Type I diabetes, especially prevalent in the younger population, to lead longer and healthier lives.

Additionally, DNA profiling has become significant in the field of forensic sciences and criminal investigations in the last few decades. Improvement in technology has allowed for faster processing and more accurate results (Arnaud, 2017) Investigators collect DNA samples from the skin cells, tissues, and/or bodily fluids of the suspect to produce a DNA profile (Arnaud, 2017). Short DNA sequences are often repeated multiple times on a chromosome at marker regions (Arnaud, 2017). The number of repeats at the marker regions varies from person to person, and investigators enter this unique DNA profile into law enforcement databases in order to determine a potential suspect (Arnaud, 2017). Nowadays, DNA profiles can be extracted from very small samples, but remain very specific (Arnaud, 2017). New data analysis techniques have made it possible for investigators to identify and distinguish multiple individuals from the DNA as well as analyze samples in less than two hours (Arnaud, 2017).

In 2003, the National Institute of Health (NIH, 2015) published results from the Human Genome Project, a 15 year project that deciphered the entire human genome, determined around 20,500 human genes and mapped their location on the chromosome (NIH, 2015). This extensive project provides detailed information about the structure, organization and function of human genes (NIH, 2015). This detailed publication acts as a sort of blueprint for the development and function of a human being, from which human traits can be tracked over generations (ref). With technologies for gene sequencing becoming more efficient and cheaper, developed for gene sequencing, human genome sequencing has expanded from academica and research to private and commercial medicine such as personal genomics services. One of the most popular of these services is 23 and Me, which allows consumers to discover their ancestry, unique traits, and learn about personal health.

CRISPR, gene editing and treating diseases

CRISPR is an innovative technology that allows for gene editing and has shown promising results in treating HIV, blood cancers and other chronic diseases (SLU, 2016). Scientists are able to use the CRISPR-Cas9 enzyme to precisely cleave DNA at target sites (SLU, 2016). Compare to previous gene editing techniques tested in human cells such as the TALE-mediated editing which had low efficiency, CRISPR-Cas9 has promised efficiency rates of up to 78% (SLU, 2016). More recent clinical applications of gene editing include correcting the mutation in the bone marrow stem cells of patients with sickle cell disease or hemophilia as well as treating schizophrenia. Since its invention, the CRISPR-Cas9 technique has been used in experimenting on lab rats, monkeys and non-viable human embryos (Kedmey, 2015). However, its application on living human embryos to 'fix' traits for rare and harmful genetic illnesses from a family's bloodline (Kedmey, 2015). The co-inventor of the CRISPR-Cas9 technology, Jennifer Doudna called for a 'pause' on gene-editing research on humans to come together as a global community and discuss ethical implications of using it to edit human genes that could potentially change human life forever (Kedmey, 2015). In 2015, Doudna and other genetic scientists participated in the International Summit on Human Gene Editing, debating the ethical concerns and setting a framework for future experimentation and implementation (Kedmey, 2015). The summit discussed many ethical and societal concerns relating to the widespread use of gene editing in improving human lives, reaching specific conclusions that are indicated below (National Academies, 2015):

- Preclinical research is needed and must proceed under appropriate legal and ethical rules.
- Within the promising clinical applications of gene editing technology, the risks and potential benefits in approving clinical trials and therapies needed to be weighed.
- Any clinical use such as germline editing should proceed only under regulatory oversight with safety and efficacy issues resolved and societal consensus about the appropriateness of the clinical use.
- There is a need for an ongoing forum and the international community should strive to establish norms around acceptable uses of human germline editing, advancing human welfare while discouraging unacceptable activities.

How research and biotechnology companies are maintaining these and new sets of guidelines, challenges and concerns that come with it, and exciting advances in this field will be discussed in the next section.

Zhou et al (2020) categorise contradictions and concerns of gene editing, mainly in embryos into three categories: security issues due to the newly developing technology, the lack of accurate definition of what classifies as necessary application and ethical use. In terms of security, concerns may arise when the sgRNAin the CRISPR/Cas9 system mismatched with the DNA fragment that it was not supposed to target, introducing unintended genetic mutation which can remain in the bloodline (Zhou et al, 2020). While the chances of these off-target effects occurring are not very high, the current techniques to fix the effect(s) require an analysis of the whole genome which can be very time-consuming and costly (Zhou et al, 2020).

Currently, Preimplantation Genetic Diagnosis (PGD) is an effective method of detecting heritable genetic diseases prior to implantation of the embryo, usually in the case of in-vitro fertilization (American Pregnancy Association). It helps select for embryos that do not carry the gene for a genetic illness. The ethics involved in PGD differ from embryo gene editing in a few ways as PDG is a method of diagnosis of any mutant gene present in the embryo rather than a treatment to correct the gene. Genetic editing has the potential to completely remove the harmful gene from passing on to future generations, but in the short run, it may be the only hope for some parents to get healthy offspring, especially when there is inevitable mitochondrial DNA mutations in the embryo.

One major and exciting prospect of gene editing is gene therapy, providing a highly effective or even a permanent treatment for common diseases such diabetes, leukemia, sickle cell anemia (SCA) and certain cancers (Zhou et al, 2020). This is achieved by introducing a normal (healthy) gene to editing and repairing defective genes. Incorporating a normal gene through somatic gene therapy may not be effective for many hereditary genetic diseases including mitochondrial diseases and Huntington's disease, and common diseases such as Alzheimer's disease and diabetes as they are regulated by many genes (Zhou et al, 2020). In 2018, the first clinical trial in the US was run, using CRISPR to treat specific cancers including melanoma. Making edits to the T cells, or cells that are responsible for immune response, recognizing invaders such as cancer cells, and killing them. However, the virus RNA can alternate the DNA of the host and there are other unidentified reasons as to why T lymphocytes can sometimes fail to recognize tumor cells or fight them effectively.

(Thulin, 2019). Within the clinical trial, patients' cells were extracted with edits made to the target genes using CRISPR. Additional genes were added on occasion to modify T cells to target cells with antigens, or chemicals produced by a foreign body. Additionally, PD-1 proteins were eliminated from T cells, the protein that can prevent T cells from targeting cancer cells, therefore increasing their efficiency (Thulin, 2019). Further studies are being conducted to study the effectiveness of cancer immunotherapy via gene editing.

Another clinical trial based in Massachusetts, U.S. aims to provide a treatment for SCA. A person with SCA may have a defect in the protein haemoglobin whereby their red blood cells are ineffective in carrying oxygen throughout the body (Thulin, 2019). The cells being irregularly shaped blood cells can't flow smoothly through blood vessels, frequently causing blockages, pain and anemia (Thulin, 2019). Stem cells are collected from the patient and edited with CRISPR so they will release higher levels of fetal hemoglobin, causing the red blood cells to sickle to a lesser degree (Thulin 2019). The trial is set to conclude in May 2022. The hopes are that it can successfully inject edited stem cells through a catheter, with about 20% fetal haemoglobin observed in patients' blood for at least three months (Thulin, 2019). This therapy could offer an effective treatment option for a disease with only a few available treatments.

Another prospect of gene editing in the field of genetic diseases is treating cystic fibrosis (CF). CF is an autosomal recessive disease that targets multiple organs. It is caused by mutation in the CFTR gene leading to dysfunction of the chloride transporter, it causes severe lung disease, chronic endocrine and exocrine pancreatic insufficiency (Mandip & Steer, 2019). The life expectancy differs is approximately 40 years but has been increasing because of various therapies (Mandip & Steer 2019, Prnake et al 2019). CFTR mutations are caused by defects at multiple stages of protein production and function (Prnake et al, 2019). Gene therapies have provided some hope for treatment by replacing mutated genes with proper copies of the CFTR gene to the epithelial cell layer in the airways (Prnake et al, 2019). However, natural barriers including mucus and immune responses could impair gene transfer into the lungs. Additionally, the airway epithelium is constantly renewing, therefore these therapies would require repeated administration (Prnake et al, 2019) The most commonly used agents in gene therapy for CF are viral vectors, and overall, the delivery of the treatment presents a barrier in its effectiveness (Prnake et al, 2019).

Through the years, CRISPR is also being examined as a viable treatment for CF. In 2015, a study by Bellec et al. (2015) successfully edited the CFTR gene in the epithelial airway cells and Calu-3 cells using the CRISPR/Cas9 approach delivered with HIV-1 lentivirus, a viral vector. The gene modification was associated with a decrease in response to a CFTR inhibitor" (Prnake et al, 2019). Pranke et al (2019) explain non-viral vectors being most efficient for CRISPR therapy as they reduce immune and limit off-target effects. More studies on the effectiveness of CRISPR-based therapy for CF are underway, with optimistic results (Prnake et al. 2019).

In the future of healthcare is an emerging field of personalized medicine to treat complex diseases including those that are inherited and genetic. This stems from knowledge and evidence that the effect of certain drugs can vary from person to person depending on their physiology, lifestyle factors, family history, and various other individual and social determinants of health (OntarioGenomics, 2014). Physicians can conduct a full genomic analysis of their patients and analyse their family history and unique lifestyle factors (OntarioGenomics, 2014).Computers can make thorough comparison with current medical data and pharmaceuticals, providing physicians with recommendations for drug and dosage for each patent (OntarioGenomics, 2014). Personalized medicine would help save on healthcare costs for patients who repeatedly purchase drugs that are ineffective or lead to lower quality of health, and save time for physicians making conclusions on otherwise, little information (OntarioGenomics, 2014). Within Ontario, Canada, personalized medicine is still in the works, with a lot of discussions and debate on the implications of incorporating personalized medicine within the existing healthcare system to better manage health resources and ensure better quality of health for Canadians (OntarioGenomics, 2014).

Designer babies and other ethical issues
Producing designer babies is a major debate in the field of biotechnology. The goal is that with designer babies, genetic and hereditary diseases can be prevented in offspring by choosing embryos that do not carry allele or gene for disease. These diseases include cystic fibrosis, cancers, b-thalassemia, Huntington's disease and other mitochondrial diseases (Pang and Ho, 2016). This can be achieved using PGD as discussed previously, however, sometimes, 100% of the embryos carry the disease of interest, thus genetic modification is a necessary option for producing healthy offspring to parents who carry the disease. Issues with using include off-target effects but also asking which kind of germline edits are safe (Pang and Ho, 2016). This requires ongoing discussion within the international community on meeting set criteria such as the 2015 International Summit on Human Gene Editing and conferences and summits that follow. In 2017, the US National Academies of Sciences,

Engineering, and Medicine outlined the conditions that should be met before editing a viable human embryo, one of which is that DNA sequences created by the edit already be common in the population, and carry no known risk of disease (Ledford 2019). This creates a barrier to treating heritable diseases through gene editing, as it is difficult to be completely certain that the variant will not increase the risk of disease (Ledford 2019). One example is of the PCSK9 gene which is associated with lower cholesterol levels and reduced risk of heart disease, and is a gene that can be potentially edited. However, only a small number of people within the population have protective mutations (Ledford, 2019).

Debate within the scientific community and public sphere asks which genes can be interfered with and which diseases can be 'fixed'. For hereditary genetic diseases such as Huntington's disease, there may be more consensus than for discussion of diseases such as heart disease and editing the PCSK9 gene (Ledford, 2019). A common area of concern is in editing genes to treat diseases versus enhancement such as improving athletic or academic capabilities, and there seems to be a blurred line in between (Ledford, 2019). There is no ethical consensus as to how to define the enhancement from prevention and moving too soon as to permit the practice could result in the creation of super humans who have an unfair genetic advantage over those unenhanced ones (Bu, pp.125). Another concern is the unequal distribution of benefits of genome editing and ensuring affordability. Since the technology is still being developed for human use, it can be assumed that it is still costly. Whether public healthcare will help to cover or alleviate some of the costs is an important question to ask. How will intellectual property rights work is also an important question to discuss when understanding the implications of social equality and fair distribution. It may be too early to tell, but these ethical concerns need to be discussed on an ongoing basis to ensure that when the gene-editing technology does become publicly available, those who need the services, especially those that are most vulnerable have adequate access to it regardless of their class, gender, sex, and race (Bu, pp.125).

A lot of studies regarding gene editing in humans are still in their preliminary stages, and many countries have decided to stay away from editing the human genome completely as of now. International panels and the scientific community reveal mixed opinions on how and if to move forward with clinical trials of human embryos (as of yet). When more preliminary research is underway, especially in gene editing for diseases such as SCA, diabetes and cancers, larger trials will have to follow before the FDA can approve any new treatment for genetic editing (Thulin, 2019). Societal consensus on the ethical use of gene editing technologies will also need to be established (National Academics, 2015). While the technology shows exciting prospects for disease treatment and therapies, many scientists argue that we would have to wait many more years before genetic editing and CRISPR therapies become readily available for those who need it within our public healthcare systems.

References

Chapter 1:

Britannica. (n.d.). Antonie an :leeuwenhoek. Retrieved from https://www.britannica.com/biography/Antonie-van-Leeuwenhoek

Charlesworth, B., & Charlesworth, D. (2009). Darwin and Genetics. Genetics, 183, 757-766.

Fentress, S. (2019, December 3). Robert Hooke's Cork Surprise. Retrieved from https://indianapublicmedia.org/amomentofscience/robert-hookes-cork-surprise.php#:~:text=of these cells.-,Robert Hooke had discovered the small-scale structure of cork,that's why cork is springy.

History Extra. (n.d.). Charles Darwin vs God: Did the 'Origin of Species' cause a clash between church and science? Retrieved from https://www.historyextra.com/period/victorian/darwin-vs-god-did-the-origin-of-species-cause-a-clash-between-church-and-science/

Johann Grego Mendel (1822-1884). (n.d.). Retrieved from http://www.dnaftb.org/1/bio.html#:~:text=Gregor Mendel, through his work,as dominant or recessive traits.

Maddox, B. (2003). Before Watson and Crick. Retrieved from https://www.pbs.org/wgbh/nova/article/before-watson-crick/

Nature Education. (n.d.). Discovery of the Function of DNA Resulted from the Work of Multiple Scientists. Retrieved from https://www.nature.com/scitable/topicpage/discovery-of-the-function-of-dna-resulted-6494318/

O'Connor, C., & Miko, I. (2008). Developing the Chromosome Theory. Retrieved from https://www.nature.com/scitable/topicpage/developing-the-chromosome-theory-164/

O'Connor, J., & Robertson, E. (2002). Robert Hooke. Retrieved from https://mathshistory.st-andrews.ac.uk/Biographies/Hooke/

Olby, R. (n.d.). Gregor Mendel. Retrieved from https://www.britannica.com/biography/Gregor-Mendel

Orgel, L. E. (1994). The Origin of Life on the Earth. Scientific American, 271(4), 76-83. Retrieved from https://www.jstor.org/stable/10.2307/24942872.
Standford Encyclopedia of Philosophy. (2019, June 19). Darwin: From Origin of Species to Descent of Man. Retrieved from https://plato.stanford.edu/entries/origin-descent/
Wakim, S., & Grewal, M. (2021, May 02). 5.2: Discovery of Cells and Cell Theory. Retrieved from https://bio.libretexts.org/Bookshelves/Human_Biology/Book:_Human_Biology_(Wakim_and_Grewal)/05:_Cells/5.02:_Discovery_of_Cells_and_Cell_Theory

Wollman, A. J., Nudd, R., Hedlund, E. G., & Leake, M. C. (2015). From Animaculum to single molecules: 300 years of the light microscope. Open Biology, 5(150019). http://dx.doi.org/10.1098/rsob.150019

Chapter 2:

Aldridge, S. (2020, January 29). The DNA story. Retrieved May 4, 2021, from https://www.chemistryworld.com/news/the-dna-story/3003946.article

Dahm, R. (2008). The First Discovery of DNA. American Scientist, 96(4), 320. doi:10.1511/2008.73.3846

Griffith, F. (1928). The Significance of Pneumococcal Types. Journal of Hygiene, 27(2), 113-159

O'Connor, C. (2008) Isolating hereditary material: Frederick Griffith, Oswald Avery, Alfred Hershey, and Martha Chase. Nature Education 1(1):105

Ogiwara, I. (2002). V-SINEs: A New Superfamily of Vertebrate SINEs That Are Widespread in Vertebrate Genomes and Retain a Strongly Conserved Segment within Each Repetitive Unit. Genome Research, 12(2), 316-324

Pray, L. (2008) Discovery of DNA structure and function: Watson and Crick. Nature Education 1(1):100

Pray, L. (2008) Transposons, or jumping genes: Not junk DNA? Nature Education 1(1):32

Rich, A., & Zhang, S. (2003). Z-DNA: The long road to biological function. Nature Reviews Genetics, 4(7), 566-572

Smit, A. F. (1999). Interspersed repeats and other mementos of transposable elements in mammalian genomes. Current Opinion in Genetics & Development, 9(6), 657-663

Chapter 3:

Abdel-Monem, M., Dürwald, H., & Hoffmann-Berling, H. (1976). Enzymic Unwinding of DNA: 2. Chain Separation by an ATP-Dependent DNA Unwinding Enzyme. European Journal of Biochemistry, 65(2), 441–449.

Abdel-Monem, M., & Hoffman-Berling, H. (1976). Enzymic Unwinding of DNA: 1. Purification and Characterization of a DNA-Dependent ATPase from Escherichia coli. European Journal of Biochemistry, 65(2), 431–440.

Baum, D. (2008). Reading a Phylogenetic Tree: The Meaning of Monophyletic Groups. Scitable. https://www.nature.com/scitable/topicpage/reading-a-phylogenetic-tree-the-meaning-of-41956/

Biro, J. C. (2008). Discovery of Proteomic Code with mRNA Assisted Protein Folding. International Journal of Molecular Sciences, 9(12), 2424–2446. https://doi.org/10.3390/ijms9122424

Brenner, S. (1957). On the Impossibility of All Overlapping Triplet Codes in Information Transfer from Nucleic Acid to Proteins. Proceedings of the National Academy of Sciences, 43(8), 687–694. https://doi.org/10.1073/pnas.43.8.687

Brimacombe, R., Trupin, J., Nirenberg, M., Leder, P., Bernfield, M., & Jaouni, T. (1965). RNA codewords and protein synthesis, 8. Nucleotide sequences of synonym codons for arginine, valine, cysteine, and alanine. Proceedings of the National Academy of Sciences of the United States of America, 54(3), 954–960.

Brown, T. A. (2002). Molecular Phylogenetics. In Genomes. 2nd edition. Wiley-Liss. https://www.ncbi.nlm.nih.gov/books/NBK21122/

Charlesworth, B., & Charlesworth, D. (2009). Darwin and Genetics. Genetics, 183(3), 757–766.

https://doi.org/10.1534/genetics.109.109991

Heidi Chial. (2008). DNA Sequencing Technologies Key to the Human Genome Project. Scitable. https://www.nature.com/scitable/topicpage/dna-sequencing-technologies-key-to-the-human-828/

Cohen, S. N., Chang, A. C., Boyer, H. W., & Helling, R. B. (1973). Construction of biologically functional bacterial plasmids in vitro. Proceedings of the National Academy of Sciences of the United States of America, 70(11), 3240–3244. https://doi.org/10.1073/pnas.70.11.3240

Crick, F. (1955). On Degenerate Templates and the Adaptor Hypothesis. https://collections.nlm.nih.gov/ext/document/101584582X73/PDF/101584582X73.pdf

Crick, F. (1966). The genetic code—Yesterday, today, and tomorrow. Cold Spring Harbor Symposia on Quantitative Biology, 31, 3–9.

Crick, F. H. C., Griffith, J. S., & Orgel, L. E. (1957). Codes Without Commas. Proceedings of the National Academy of Sciences of the United States of America, 43(5), 416–421.

Dahm, R. (2005). Friedrich Miescher and the discovery of DNA. Developmental Biology, 278(2), 274–288. https://doi.org/10.1016/j.ydbio.2004.11.028

Danna, K., & Nathans, D. (1971). Specific Cleavage of Simian Virus 40 DNA by Restriction Endonuclease of Hemophilus Influenzae*. Proceedings of the National Academy of Sciences of the United States of America, 68(12), 2913–2917.

Forsdyke, D. R. (2010). The Selfish Gene Revisited: Reconciliation of Williams-Dawkins and Conventional Definitions. Biological Theory, 5(3), 246–255. https://doi.org/10.1162/BIOT_a_00054

Friedberg, E. C. (2006). The eureka enzyme: The discovery of DNA polymerase. Nature Reviews Molecular Cell Biology, 7(2), 143–147. https://doi.org/10.1038/nrm1787

Geider, K. (1978). The single-stranded DNA phages.(eds. Denhardt, Dressler & Ray) 378–387. Cold Spring Habor Laboratory Press, New York.

George Gamow. (1953). Letter from George Gamow to Linus Pauling. October 22, 1953. Transcript—Correspondence—Linus Pauling and the Race for DNA: A Documentary History. http://scarc.library.oregonstate.edu/coll/pauling/dna/corr/sci9.001.43-gamow-lp-19531022-transcript.html

Goldstein, B. (2018). On Francis Crick, the genetic code, and a clever kid. Current Biology, 28(7), R305. https://doi.org/10.1016/j.cub.2018.02.058

Heather, J. M., & Chain, B. (2016). The sequence of sequencers: The history of sequencing DNA. Genomics, 107(1), 1–8. https://doi.org/10.1016/j.ygeno.2015.11.003

Hellmich, R. L. (2012). Use and Impact of Bt Maize. Scitable. https://www.nature.com/scitable/knowledge/library/use-and-impact-of-bt-maize-46975413/

Huebscher, U. (Ed.). (1984). Proteins Involved in DNA Replication. Springer US. https://doi.org/10.1007/978-1-4684-8730-5

Kornberg, A., Lehman, I. R., & Simms, E. S. (1956). Polydesoxyribonucleotide synthesis by enzymes from Escherichia coli. Fed. Proc, 15, 291–292.

Kornberg, T., & Gefter, M. L. (1970). DNA synthesis in cell-free extracts of a DNA polymerase-defective mutant. Biochemical and Biophysical Research Communications, 40(6), 1348–1355.

Lander, E. S., Linton, L. M., Birren, B., Nusbaum, C., Zody, M. C., Baldwin, J., Devon, K., Dewar, K., Doyle, M., FitzHugh, W., Funke, R., Gage, D., Harris, K., Heaford, A., Howland, J., Kann, L., Lehoczky, J., LeVine, R., McEwan, P., ... The Wellcome Trust: (2001). Initial sequencing and analysis of the human genome. Nature, 409(6822), 860–921. https://doi.org/10.1038/35057062

Lehman, I. R., Bessman, M. J., Simms, E. S., & Kornberg, A. (1958). Enzymatic synthesis of deoxyribonucleic acid: V. Chemical composition of enzymatically synthesized deoxyribonucleic acid. Proc. Natl. Acad. Sci. USA, 1191–1196.

Lehman, I. R., Zimmerman, S. B., Adler, J., Bessman, M. J., Simms, E. S., & Kornberg, A. (1958). Enzymatic synthesis of deoxyribonucleic acid. V. Chemical composition of enzymatically synthesized deoxyribonucleic acid. Proceedings of the National Academy of Sciences of the United States of America, 44(12), 1191.

Li, H., Yang, Y., Hong, W., Huang, M., Wu, M., & Zhao, X. (2020). Applications of genome editing technology in the targeted therapy of human diseases: Mechanisms, advances and prospects. Signal Transduction and Targeted Therapy, 5(1), 1–23. https://doi.org/10.1038/s41392-019-0089-y

Matthaei, J. H., & Nirenberg, M. W. (1961). Characteristics and stabilization of DNAase-sensitive protein synthesis in E. coli extracts. Proceedings of the National Academy of Sciences of the United States of America, 47, 1580–1588. https://doi.org/10.1073/pnas.47.10.1580

Meselson, M., & Stahl, F. W. (1958). The replication of DNA in Escherichia coli. Proceedings of the National Academy of Sciences, 44(7), 671–682.
Meselson, M., & Yuan, R. (1968). DNA Restriction Enzyme from E. coli. Nature, 217(5134), 1110–1114. https://doi.org/10.1038/2171110a0

Miko, I. (2009). Gregor Mendel and the Principles of Inheritance. Scitable. https://www.nature.com/scitable/topicpage/gregor-mendel-and-the-principles-of-inheritance-593/

Mullis, K. B., & Faloona, F. A. (1987). Specific synthesis of DNA in vitro via a polymerase-catalyzed chain reaction. Methods in Enzymology, 155, 335–350. https://doi.org/10.1016/0076-6879(87)55023-6

Mullis, Kary B., Ferré, F., & Gibbs, R. A. (Eds.). (1994). The Polymerase Chain Reaction. Birkhäuser Boston. https://doi.org/10.1007/978-1-4612-0257-8

Nanjundiah, V. (2004). George Gamow and the genetic code. Resonance, 9(7), 44–49. https://doi.org/10.1007/BF02903575

Nirenberg, M. (2004). Historical review: Deciphering the genetic code – a personal account. Trends in Biochemical Sciences, 29(1), 46–54. https://doi.org/10.1016/j.tibs.2003.11.009

Nirenberg, M. W., & Matthaei, J. H. (1961). The dependence of cell-free protein synthesis in E. coli upon naturally occurring or synthetic polyribonucleotides. Proceedings of the National Academy of Sciences, 47(10), 1588–1602. https://doi.org/10.1073/pnas.47.10.1588

Palca, J. (1986). Human genome: Department of Energy on the map. Nature, 321(6068), 371–371. https://doi.org/10.1038/321371a0

Phillips, T. (2008). Genetically Modified Organisms (GMOs): Transgenic Crops and Recombinant DNA Technology. Scitable. https://www.nature.com/scitable/topicpage/discovery-of-dna-structure-and-function-watson-397/

Pray, L. A. (2008). Discovery of DNA Structure and Function: Watson and Crick. Scitable. https://www.nature.com/scitable/topicpage/discovery-of-dna-structure-and-function-watson-397/

Rich, A. (1997). Gamow and the genetic Code. 114–122.

Roberts, R. J. (2005). How restriction enzymes became the workhorses of molecular biology. Proceedings of the National Academy of Sciences, 102(17), 5905–5908. https://doi.org/10.1073/pnas.0500923102

Saiki, R. K., Gelfand, D. H., Stoffel, S., Scharf, S. J., Higuchi, R., Horn, G. T., Mullis, K. B., & Erlich, H. A. (1988). Primer-directed enzymatic amplification of DNA with a thermostable DNA polymerase. Science, 239(4839), 487–491. https://doi.org/10.1126/science.239.4839.487

Sanger, F., Nicklen, S., & Coulson, A. R. (1977). DNA sequencing with chain-terminating inhibitors. Proceedings of the National Academy of Sciences, 74(12), 5463–5467.

Sorsby, A. (1965). Gregor Mendel. British Medical Journal, 1(5431), 333–338.
Thömmes, P., & Hübscher, U. (1990). Eukaryotic DNA replication. European Journal of Biochemistry, 194(3), 699–712. https://doi.org/10.1111/j.1432-1033.1990.tb19460.x

Watson, J. D., & Crick, F. H. C. (1953a). Molecular Structure of Nucleic Acids: A Structure for Deoxyribose Nucleic Acid. Nature, 171(4356), 737–738. https://doi.org/10.1038/171737a0

Watson, J. D., & Crick, F. H. C. (1953b). Genetical Implications of the Structure of Deoxyribonucleic Acid. Nature, 171(4361), 964–967. https://doi.org/10.1038/171964b0

Woese, C. R. (1969). The Biological Significance of the Genetic Code. In B.

W. Agranoff, J. Davies, F. E. Hahn, H. G. Mandel, N. S. Scott, R. M. Smillie, C. R. Woese, & F. E. Hahn (Eds.), Progress in Molecular and Subcellular Biology (pp. 5–46). Springer. https://doi.org/10.1007/978-3-642-46200-9_2
World Health Organization (Ed.). (2002). Genomics and world health: Report of the Advisory Committee on Health Research. World Health Organization.

Chapter 4:
Adrio, J.-L., & Demain, A. L. (2010). Recombinant organisms for production of industrial products. Bioengineered Bugs, 1(2), 116–131. https://doi.org/10.4161/bbug.1.2.10484

Baeshen, N. A., Baeshen, M. N., Sheikh, A., Bora, R. S., Ahmed, M. M. M., Ramadan, H. A. I., Saini, K. S., & Redwan, E. M. (2014). Cell factories for insulin production. Microbial Cell Factories, 13. https://doi.org/10.1186/s12934-014-0141-0

Brock, J.-A. K., Allen, V. M., Kieser, K., & Langlois, S. (2010). Family History Screening: Use of the Three Generation Pedigree in Clinical Practice. Journal of Obstetrics and Gynaecology Canada, 32(7), 663–672. https://doi.org/10.1016/S1701-2163(16)34570-4

Canadian Cancer Society. (n.d.). Personalized medicine. Www.Cancer.Ca. Retrieved May 7, 2021, from https://www.cancer.ca:443/en/cancer-information/cancer-101/cancer-research/personalized-medicine/?region=on

Clancy, S. (2008). DNA Transcription. Nature Education 1(1):41.

D'Andrea, A. D. (2015). 4—DNA Repair Pathways and Human Cancer. In J. Mendelsohn, J. W. Gray, P. M. Howley, M. A. Israel, & C. B. Thompson (Eds.), The Molecular Basis of Cancer (Fourth Edition) (pp. 47-66.e2). W.B. Saunders. https://doi.org/10.1016/B978-1-4557-4066-6.00004-4

Doolittle, W. F. (1999). Phylogenetic classification and the universal tree. Science (New York, N.Y.), 284(5423), 2124–2129. https://doi.org/10.1126/science.284.5423.2124

Garlinghouse, T. (2019, December 18). Origin story: Rewriting human history through our DNA. Princeton University. https://www.princeton.edu/news/2019/12/18/origin-story-rewriting-human-history-through-our-dna

DNA Profiling: How Is It Used in Criminal Justice? (2019, December 31). Maryville University. https://online.maryville.edu/blog/how-is-dna-profiling-used-to-solve-crimes/

Genetic Alliance, & District of Columbia Department of Health. (2010). Diagnosis of a Genetic Disease. In Understanding Genetics: A District of Columbia Guide for Patients and Health Professionals. Genetic Alliance. https://www.ncbi.nlm.nih.gov/books/NBK132142/

Jonsson, H., Magnusdottir, E., Eggertsson, H. P., Stefansson, O. A., Arnadottir, G. A., Eiriksson, O., Zink, F., Helgason, E. A., Jonsdottir, I., Gylfason, A., Jonasdottir, A., Jonasdottir, A., Beyter, D., Steingrimsdottir, T., Norddahl, G. L., Magnusson, O. T., Masson, G., Halldorsson, B. V., Thorsteinsdottir, U., ... Stefansson, K. (2021). Differences between germline genomes of monozygotic twins. Nature Genetics, 53(1), 27–34. https://doi.org/10.1038/s41588-020-00755-1

Key, S., Ma, J. K.-C., & Drake, P. M. (2008). Genetically modified plants and human health. Journal of the Royal Society of Medicine, 101(6), 290–298. https://doi.org/10.1258/jrsm.2008.070372

Liang, P., Xu, Y., Zhang, X., Ding, C., Huang, R., Zhang, Z., Lv, J., Xie, X., Chen, Y., Li, Y., Sun, Y., Bai, Y., Songyang, Z., Ma, W., Zhou, C., & Huang, J. (2015). CRISPR/Cas9-mediated gene editing in human tripronuclear zygotes. Protein & Cell, 6(5), 363–372. https://doi.org/10.1007/s13238-015-0153-5

Lee, P. Y., Costumbrado, J., Hsu, C.-Y., & Kim, Y. H. (2012). Agarose Gel Electrophoresis for the Separation of DNA Fragments. Journal of Visualized Experiments, 62, e3923. https://doi.org/10.3791/3923

Miller, W. L., & Baxter, J. D. (1980). Recombinant DNA — A new source of insulin. Diabetologia, 18(6), 431–436. https://doi.org/10.1007/BF00261696

Patel, P. H., & Zulfiqar, H. (2021). Reverse Transcriptase Inhibitors. In StatPearls. StatPearls

Publishing. http://www.ncbi.nlm.nih.gov/books/NBK551504/

Pardi, N., Hogan, M. J., Porter, F. W., & Weissman, D. (2018). mRNA vaccines—A new era in vaccinology. Nature Reviews Drug Discovery, 17(4), 261–279. https://doi.org/10.1038/nrd.2017.243

Public Health Agency of Canada. (2013, February 5). Genetic testing and screening [Education and awareness]. Government of Canada. https://www.canada.ca/en/public-health/services/fertility/genetic-testing-screening.html

Ralhan, R., & Kaur, J. (2007). Alkylating agents and cancer therapy. Expert Opinion on Therapeutic Patents, 17(9), 1061–1075. https://doi.org/10.1517/13543776.17.9.1061

Vogenberg, F. R., Isaacson Barash, C., & Pursel, M. (2010). Personalized Medicine. Pharmacy and Therapeutics, 35(10), 560–576.

World Health Organization. (n.d.). Food, Genetically modified. Retrieved May 7, 2021, from https://www.who.int/westernpacific/health-topics/food-genetically-modified

Chapter 5:

Liu, Y. S., Zhou, X. M., Zhi, M. X., Li, X. J., & Wang, Q. L. (2009). Darwin's contributions to genetics. Journal of applied genetics, 50(3), 177–184. https://doi.org/10.1007/BF03195671

Liu Y. (2018). The Influence of Darwin's Pangenesis on Later Theories. Advances in genetics, 101, 63–85. https://doi.org/10.1016/bs.adgen.2018.05.003

Sandler, I. (2000). Development: Mendel's Legacy to Genetics. Genetics, 154(1), 7–11. https://doi.org/10.1093/genetics/154.1.7

Gliboff, S. (2015). The Mendelian and Non-Mendelian Origins of Genetics. Filosofia e História Da Biologia, 10(1), 99–123.

Dahm R. (2008). Discovering DNA: Friedrich Miescher and the early years of nucleic acid research. Human genetics, 122(6), 565–581. https://doi.org/10.1007/s00439-007-0433-0

KLUG, A. (1968). Rosalind Franklin and the Discovery of the Structure of DNA. Nature, 219(5156), 808–810. https://doi.org/10.1038/219808a0

Smyth, M. S., & Martin, J. H. (2000). x ray crystallography. Molecular pathology : MP, 53(1), 8–14. https://doi.org/10.1136/mp.53.1.8

Egli, M., Tereshko, V., Teplova, M., Minasov, G., Joachimiak, A., Sanishvili, R., Weeks, C.M., Miller, R., Maier, M.A., An, H., Dan Cook, P. and Manoharan, M. (1998), X-ray crystallographic analysis of the hydration of A- and B-form DNA at atomic resolution. Biopolymers, 48: 234-252. https://doi.org/10.1002/(SICI)1097-0282(1998)48:4<234::AID-BIP4>3.0.CO;2-H

National Academies Press. (2017). Human genome editing: science, ethics, and governance.

Ma, Y., Zhang, L. and Huang, X. (2014), Genome modification by CRISPR/Cas9. FEBS J, 281: 5186-5193. https://doi.org/10.1111/febs.13110

National Institutes of Health. (2020a, August 17). Polymerase Chain Reaction (PCR) Fact Sheet. Genome.gov. https://www.genome.gov/about-genomics/fact-sheets/Polymerase-Chain-Reaction-Fact-Sheet.

Alonso, A., Martın, P., Albarrán, C., Garcia, P., Garcia, O., Fernández, L. de S., Garcia-Hirschfeld J., Sancho M., de la Rúa C., F Fernández-Piqueras, J. (2004). Real-time PCR designs to estimate nuclear and mitochondrial DNA copy number in forensic and ancient DNA studies. Forensic Science International, 139(2-3), 141–149. https://doi.org/10.1016/j.forsciint.2003.10.008

Shao, M., Xu, T. R., & Chen, C. S. (2016). The big bang of genome editing technology: development and application of the CRISPR/Cas9 system in disease animal models. Dong wu xue yan jiu = Zoological research, 37(4), 191–204. https://doi.org/10.13918/j.issn.2095-8137.2016.4.191

Go, D. E., & Stottmann, R. W. (2016). The Impact of CRISPR/Cas9-Based Genomic Engineering on Biomedical Research and Medicine. Current molecular medicine, 16(4), 343–352. https://doi.org/10.2174/1566524016666160316150847

Lumpkin, O. J., Déjardin, P., & Zimm, B. H. (1985). Theory of gel electrophoresis of DNA. Biopolymers, 24(8), 1573–1593. https://doi.org/10.1002/bip.360240812

Aaij, C., & Borst, P. (1972). The gel electrophoresis of DNA. Biochimica Et Biophysica Acta (BBA) - Nucleic Acids and Protein Synthesis, 269(2), 192–200. https://doi.org/10.1016/0005-2787(72)90426-1

National Institutes of Health. (2020b). DNA Microarray Technology Fact Sheet. Genome.gov. https://www.genome.gov/about-genomics/fact-sheets/DNA-Microarray-Technology.

Heller, M. J. (2002). DNA Microarray Technology: Devices, Systems, and Applications. Annual Review of Biomedical Engineering, 4(1), 129–153. https://doi.org/10.1146/annurev.bioeng.4.020702.153438

National Institutes of Health. (2020c, February 24). Human Genome Project FAQ. Genome.gov. https://www.genome.gov/human-genome-project/Completion-FAQ.

Behjati, S., & Tarpey, P. S. (2013). What is next generation sequencing?. Archives of disease in childhood. Education and practice edition, 98(6), 236–238. https://doi.org/10.1136/archdischild-2013-304340

Iranbakhsh, A., & Seyyedrezaei, S. H. (2011). The impact of information technology in biological sciences. Procedia Computer Science, 3, 913–916. https://doi.org/10.1016/j.procs.2010.12.149

Senior, A. W., Evans, R., Jumper, J., Kirkpatrick, J., Sifre, L., Green, T., Qin C., Žídek A., Nelson A., Bridgland A., Penedones H., Petersen S., Simonyan K., Crossan S., Kohli P., Jones D., Silver D., Kavukcuoglu K., Hassabis, D. (2020). Improved protein structure prediction using potentials from deep learning. Nature, 577(7792), 706–710. https://doi.org/10.1038/s41586-019-1923-7

Farid F. Abraham (1986) Computational statistical mechanics methodology, applications and supercomputing, Advances in Physics, 35:1, 1-111, DOI: 10.1080/00018738600101851

Nicholas, H., Giras, G., Hartonas-Garmhausen, V., Kopko, M., Maher, C., & Ropelewski, A. (1991). Distributing the comparison of DNA and protein sequences across heterogeneous supercomputers. Proceedings of the 1991 ACM/IEEE Conference on Supercomputing - Supercomputing '91. https://doi.org/10.1145/125826.125911

Parducci, L. (2019). Quaternary DNA: A Multidisciplinary Research Field. Quaternary, 2(4), 37. https://doi.org/10.3390/quat2040037

Cipollaro, M., Galderisi, U., & Di Bernardo, G. (2004). Ancient DNA as a multidisciplinary experience. Journal of Cellular Physiology, 202(2), 315–322. https://doi.org/10.1002/jcp.20116

Chapter 6:
Biskup, S., & Gasser, T. (2012). Genetic testing in neurological diseases. Journal of Neurology, 259(6), 1249-1254. https://doi.org/10.1007/s00415-012-6511-9

Blashki, G., Metcalfe, S., & Emery, J. (2014). Genetics in general practice. Genetics, 43(7), 428-431. https://www.racgp.org.au/afp/2014/july/genetics-in-general-practice/

Bowers, W. J., Breakefiel d, X. O., & Sena-Esteves, M. (2011). Genetic therapy for the nervous system. Human Molecular Genetics, 20(1), 28-41. https://doi-org.cyber.usask.ca/10.1093/hmg/ddr110

Carreno, B. M., Magrini, V., Becker-Hapak, M., Kaabinejadian, S., Hundal, J., Petti, A. A., Ly, A., Lie, W., Hildebrand, W. H., Mardis, E. R., & Linette, G. P. (2015). A dendritic cell vaccine increases the breadth and diversity of melanoma neoantigen-specific T cells. Science, 348(6236), 803-808. https://doi.org/10.1126/science.aaa3828

Heid, M. (2015). DNA-based personalized medicine may change healthcare forever. Shape. https://www.shape.com/lifestyle/mind-and-body/dna-based-personalized-medicine-may-change-healthcare-forever

Kaufman, E. S. (2012). Genetic testing in Brugada syndrome. Journal of the American College of Cardiology, 60(15), 1419-1420. https://www-clinicalkey-com.cyber.usask.ca/#!/content/playContent/1-s2.0-S0735109712019481?returnurl=null&referrer=null

MayoClinic. (2021). Center for individualized medicine - Pharmacogenomics in patient care. https://www.mayo.edu/research/centers-programs/center-individualized-medicine/patient-care/pharmacogenomics

Ooi, C. Y., Gonska, T., Durie, P. R., & Freedman, S. D. (2010). Genetic testing in pancreatitis. Gastroenterology, 138(7), 2202-2206.e1. https://doi.org/10.1053/j.gastro.2010.04.022
Ross, P. T. (2015). Motivations of women with sickle cell disease for asking their partners to undergo genetic testing. Social Science & Medicine, 139, 36-43. https://doi.org/10.1016/j.socscimed.2015.06.029

Wilkinson, G. W., & Borysiewicz, L. K. (1995). Gene therapy and viral vaccination: the interface. British Medical Bulletin, 51(1), 205-216. https://doi.org/10.1097/00013542-199404000-00003

Chapter 7:
Ahn, J. & Lee, J. (2008) X chromosome: X inactivation. Nature Education 1(1):24

Bagchi, A. (2020). Unusual nature of long non-coding RNAs coding for "unusual peptides." Gene, 729, 144298. https://doi.org/10.1016/j.gene.2019.144298

Ball, P. Celebrate the unknowns. Nature 496, 419–420 (2013). https://doi.org/10.1038/496419a

Chial, H. (2008) DNA sequencing technologies key to the Human Genome Project. Nature Education 1(1):219

Crouch, D. J. M., & Bodmer, W. F. (2020). Polygenic inheritance, GWAS, polygenic risk scores, and the search for functional variants. Proceedings of the National Academy of Sciences, 117(32), 18924–18933. https://doi.org/10.1073/pnas.2005634117

Díaz-Castillo, C. (2017). Junk DNA Contribution to Evolutionary Capacitance Can Drive Species Dynamics. Evolutionary Biology, 44(2), 190–205. https://doi.org/10.1007/s11692-016-9404-5

Doolittle, W. F. (2013). Is junk DNA bunk? A critique of ENCODE. Proceedings of the National Academy of Sciences, 110(14), 5294–5300. https://doi.org/10.1073/pnas.1221376110

Dunham, I., Kundaje, A., Aldred, S. F., Collins, P. J., Davis, C. A., Doyle, F., Epstein, C. B., Frietze, S., Harrow, J., Kaul, R., Khatun, J., Lajoie, B. R., Landt, S. G., Lee, B.-K., Pauli, F., Rosenbloom, K. R., Sabo, P., Safi, A., Sanyal, A., … HudsonAlpha Institute, C., UC Irvine, Stanford group (data production and analysis). (2012). An integrated encyclopedia of DNA elements in the human genome. Nature, 489(7414), 57–74. https://doi.org/10.1038/nature11247

Evolution—Latest research and news | Nature. (2021). Springer Nature Limited. https://www.nature.com/subjects/evolution

Explainer: Peptides vs proteins - what's the difference? (2017, November 13). University of Queensland. https://imb.uq.edu.au/article/2017/11/explainer-peptides-vs-proteins-whats-difference

Feigin, M. E., Garvin, T., Bailey, P., Waddell, N., Chang, D. K., Kelley, D. R., Shuai, S., Gallinger, S., McPherson, J. D., Grimmond, S. M., Khurana, E., Stein, L. D., Biankin, A. V., Schatz, M. C., & Tuveson, D. A. (2017). Recurrent noncoding regulatory mutations in pancreatic ductal adenocarcinoma.

Nature Genetics, 49(6), 825–833. https://doi.org/10.1038/ng.3861

Gene Expression | Learn Science at Scitable. (2014). Nature Education. https://www.nature.com/scitable/topicpage/gene-expression-14121669/

Graur, D. (2017). An Upper Limit on the Functional Fraction of the Human Genome. Genome Biology and Evolution, 9(7), 1880–1885. https://doi.org/10.1093/gbe/evx121

Kim, U. K., & Drayna, D. (2005). Genetics of individual differences in bitter taste perception: Lessons from the PTC gene. Clinical Genetics, 67(4), 275–280. https://doi.org/10.1111/j.1399-0004.2004.00361.x

Ling, H., Vincent, K., Pichler, M., Fodde, R., Berindan - Neagoe, I., Slack, F., & Calin, G. (2015). Junk DNA and the long non-coding RNA twist in cancer genetics. Oncogene, 34. https://doi.org/10.1038/onc.2014.456

Lobo, I. (2008) Pleiotropy: One Gene Can Affect Multiple Traits. Nature Education 1(1):10

Masotti, M., Guo, B., & Wu, B. (2019). Pleiotropy informed adaptive association test of multiple traits using genome-wide association study summary data. Biometrics, 75(4), 1076–1085.

Merritt, R. B., Bierwert, L. A., Slatko, B., Weiner, M. P., Ingram, J., Sciarra, K., & Weiner, E. (2008). Tasting Phenylthiocarbamide (PTC): A New Integrative Genetics Lab with an Old Flavor. The American Biology Teacher, 70(5). https://doi.org/10.1662/0002-7685(2008)70[23:TPPANI]2.0.CO;2

Ohno, S. (1972). So much "junk" DNA in our genome. Brookhaven Symposia in Biology, 23, 366–370.

Pleiotropy informed adaptive association test of multiple traits using genome-wide association study summary data—Masotti—2019—Biometrics—Wiley Online Library. (n.d.). Retrieved May 7, 2021, from https://onlinelibrary-wiley-com.myaccess.library.utoronto.ca/doi/full/10.1111/biom.13076

Single-Gene Disorders. (2010). In Understanding Genetics: A District of Columbia Guide for Patients and Health Professionals. Genetic Alliance. https://www.ncbi.nlm.nih.gov/books/NBK132154/

What is DNA?: MedlinePlus Genetics. (2021, January 19). MedlinePlus Genetics. https://medlineplus.gov/genetics/understanding/basics/dna/

Wei, J.-W., Huang, K., Yang, C., & Kang, C.-S. (2017). Non-coding RNAs as regulators in epigenetics (Review). Oncology Reports, 37(1), 3–9. https://doi.org/10.3892/or.2016.5236

Wood, V., Lock, A., Harris, M. A., Rutherford, K., Bähler, J., & Oliver, S. G. (2019). Hidden in plain sight: What remains to be discovered in the eukaryotic proteome? Open Biology, 9(2), 180–241. https://doi.org/10.1098/rsob.180241

Zhou, J., Park, C. Y., Theesfeld, C. L., Wong, A. K., Yuan, Y., Scheckel, C., Fak, J. J., Funk, J., Yao, K., Tajima, Y., Packer, A., Darnell, R. B., & Troyanskaya, O. G. (2019). Whole-genome deep-learning analysis identifies contribution of noncoding mutations to autism risk. Nature Genetics, 51(6), 973–980. https://doi.org/10.1038/s41588-019-0420-0

Chapter 8:

Advancing justice through DNA technology: using DNA to solve crimes(2021). Retrieved 7 May 2021, from https://www.justice.gov/archives/ag/advancing-justice-through-dna-technology-using-dna-solve-crimes

Anderson, W. (1985). Human Gene Therapy: Scientific and Ethical Considerations. Journal Of Medicine
And Philosophy, 10(3), 275-292. doi: 10.1093/jmp/10.3.275

Arnaud, C. (2021). Retrieved 7 May 2021, from
https://cen.acs.org/analytical-chemistry/Thirty-years-DNA-forensics-DNA/95/i37

Arnaud, C. (2021). Retrieved 7 May 2021, from
https://cen.acs.org/analytical-chemistry/Thirty-years-DNA-forensics-DNA/95/i37
Canadian Survey on Disability - Reports A demographic, employment and income profile of Canadians with disabilities aged 15 years and over, 2017 | LDAC-ACTA. (2021). Retrieved 8 May 2021, from https://www.ldac-acta.ca/canadian-survey-on-disability-reports-a-demographic-employment-and-income-profile-of-canadians-with-disabilities-aged-15-years-and-over-2017/#:~:text=In%202017%2C%20one%20in%20five,aged%2075%20years%20and%20over.

Steps to Justice. (2021). Retrieved 7 May 2021, from
https://stepstojustice.ca/questions/criminal-law/do-i-have-give-bodily-sample-police

Dror, I., & Hampikian, G. (2011). Subjectivity and bias in forensic DNA mixture interpretation. Science &
Justice, 51(4), 204-208. doi: 10.1016/j.scijus.2011.08.004

Ferriman, A. (2001). First cases of human germline genetic modification announced. BMJ, 322(7295), 1144-1144. doi: 10.1136/bmj.322.7295.1144

Germ Line Gene Therapy - an overview | ScienceDirect Topics. (2021). Retrieved 8 May 2021, from
https://www.sciencedirect.com/topics/medicine-and-dentistry/germ-line-gene-therapy#:~:text=In 20germline%20gene%20therapy%2C%20DNA,passed%20down%20to%20future%20generations.

Harvard researchers share views on future, ethics of gene editing. (2021). Retrieved 8 May 2021, from https://news.harvard.edu/gazette/story/2019/01/perspectives-on-gene-editing/

Thirty years of DNA forensics: How DNA has revolutionized criminal investigations. (2021). Retrieved 7 May 2021, from https://cen.acs.org/analytical-chemistry/Thirty-years-DNA-forensics-DNA/95/i37

Is germline gene therapy ethical?. (2021). Retrieved 8 May 2021, from
https://www.yourgenome.org/debates/is-germline-gene-therapy-ethical

Kleiderman, E., & Stedman, I. (2019). Human germline genome editing is illegal in Canada, but could it be desirable for some members of the rare disease community?. Journal Of Community Genetics, 11(2), 129-138. doi: 10.1007/s12687-019-00430-x

Leahy, S. (2021). Golden State Killer Busted by DNA. But Are Tests 100% Accurate?. Retrieved 7 May
2021, from
https://www.nationalgeographic.com/science/article/dna-testing-accuracy-golden-state-killer-science-spd#:~:text=The%20more%20markers%20used%2C%20the,than%201%20in%20 1%20billion.&text=Even%20advanced%20DNA%20testing%2C%20which,got%20to%20a%20 crime%20scene.

Pray, L. (2008) Discovery of DNA structure and function: Watson and Crick. Nature Education1(1):100

Shaer, M. (2016). The False Promise of DNA Testing. Retrieved 7 May 2021, from
    https://www.theatlantic.com/magazine/archive/2016/06/a-reasonable-doubt/480747/

Chapter 9:

E. coli – the biotech bacterium. Science Learning Hub. (n.d.). https://www.sciencelearn.org.nz/resources/1899-e-coli-the-biotech-bacterium#:~:text=It%20grows%20fast.,cell%20within%20 about%207%20hours.

Hernandez, V. (2017, September 21). The Embryo Project Encyclopedia. "On the Replication of Deoxyribonucleic Acid (DNA)" (1954), by Max Delbruck | The Embryo Project Encyclopedia.

https://embryo.asu.edu/pages/replication-desoxyribonucleic-acid-dna-1954-max-delbruck. How DNA Replicates. X Bio. (n.d.). https://explorebiology.org/collections/genetics/how-dna-replicates.

Khan Academy. (n.d.). Mode of DNA replication: Meselson-Stahl experiment (article). Khan Academy. https://www.khanacademy.org/science/biology/dna-as-the-genetic-material/dna-replication/a/mode-of-dna-replication-meselson-stahl-experiment.

Portin, P. (2014). The birth and development of the DNA theory of inheritance: sixty years since the discovery of the structure of DNA. Journal of Genetics, 93(1), 293–302. https://doi.org/10.1007/s12041-014-0337-4

The Nature of Science. Climate Science Investigations South Florida - The Nature of Science. (n.d.). http://www.ces.fau.edu/nasa/introduction/scientific-inquiry/why-must-scientists-be-skeptics.php#:~:text=Why%20is%20maintaining%20a%20skeptical,evidence%20to%20back%20them%20up.

The Nobel Prize in Physiology or Medicine 1969. NobelPrize.org. (n.d.). https://www.nobelprize.org/prizes/medicine/1969/delbruck/facts/.

Chapter 10:

Acuna-Hidalgo, R., Veltman, J. A., & Hoischen, A. (2016). New insights into the generation and role of de novo mutations in health and disease. Genome Biology, 17(1), 241. https://doi.org/10.1186/s13059-016-1110-1

Ball, P. (2020). The lightning-fast quest for COVID vaccines—And what it means for other diseases. Nature, 589(7840), 16–18. https://doi.org/10.1038/d41586-020-03626-1

Bijanzadeh, M., Mahesh, P. A., & Ramachandra, N. B. (2011). An understanding of the genetic basis of asthma. The Indian Journal of Medical Research, 134(2), 149–161.

Douglas, K. M., Sutton, R. M., & Cichocka, A. (2017). The Psychology of Conspiracy Theories. Current Directions in Psychological Science, 26(6), 538–542. https://doi.org/10.1177/0963721417718261

Herrera-Estrella, L., & Alvarez-Morales, A. (2001). Genetically modified crops: Hope for developing countries? EMBO Reports, 2(4), 256–258. https://doi.org/10.1093/embo-reports/kve075

Hodgson, S. (2008). Mechanisms of inherited cancer susceptibility. Journal of Zhejiang University. Science. B, 9(1), 1–4. https://doi.org/10.1631/jzus.B073001

Hsu, P. D., Lander, E. S., & Zhang, F. (2014). Development and Applications of CRISPR-Cas9 for Genome Engineering. Cell, 157(6), 1262–1278. https://doi.org/10.1016/j.cell.2014.05.010

Husaini, A. M., & Sohail, M. (2018). Time to Redefine Organic Agriculture: Can't GM Crops Be Certified as Organics? Frontiers in Plant Science, 9. https://doi.org/10.3389/fpls.2018.00423

Levine, H. (2020, September 23). The 5 Stages of COVID-19 Vaccine Development: What You Need to Know About How a Clinical Trial Works. Content Lab U.S. https://www.jnj.com/innovation/the-5-stages-of-covid-19-vaccine-development-what-you-need-to-know-about-how-a-clinical-trial-works

Mcmaster, G. (2018). Indigenous DNA no proof of Indigenous identity, says Native studies scholar. https://www.ualberta.ca/folio/2018/11/indigenous-dna-no-proof-of-indigenous-identity-says-native-studies-scholar.html

Mohsen, H. (2020). Race and Genetics: Somber History, Troubled Present. The Yale Journal of Biology and Medicine, 93(1), 215–219.

Norris. (2015, August 10). Will GMOs Hurt My Body? The Public's Concerns and How Scientists Have Addressed Them. Science in the News. https://sitn.hms.harvard.edu/flash/2015/will-gmos-hurt-my-body/

Oliver, M. J. (2014). Why We Need GMO Crops in Agriculture. Missouri Medicine, 111(6), 492–507.

Panofsky, A., & Donovan, J. (2019). Genetic ancestry testing among white nationalists: From identity repair to citizen science. Social Studies of Science, 49(5), 653–681. https://doi.org/10.1177/0306312719861434

Prakash, C. (2015). GM crops in the media. GM Crops & Food, 6(2), 63–68. https://doi.org/10.1080/21645698.2015.1056680

Pulendran, B., & Ahmed, R. (2011). Immunological mechanisms of vaccination. Nature Immunology, 12(6), 509–517.

Sachs, H. (2019). Perspective | The dark side of our genealogy craze. Washington Post. https://www.washingtonpost.com/outlook/2019/12/13/dark-side-our-genealogy-craze/

Schwarcz, J. (2017, March 20). There are No Fish Genes in Tomatoes. Office for Science and Society. https://www.mcgill.ca/oss/article/controversial-science-environment-food-health-news/there-are-no-fish-genes-tomatoes

Swire-Thompson, B., & Lazer, D. (2020). Public Health and Online Misinformation: Challenges and Recommendations. Annual Review of Public Health, 41(1), 433–451. https://doi.org/10.1146/annurev-publhealth-040119-094127

Tsaousis, G. N., Papadopoulou, E., Apessos, A., Agiannitopoulos, K., Pepe, G., Kampouri, S., Diamantopoulos, N., Floros, T., Iosifidou, R., Katopodi, O., Koumarianou, A., Markopoulos, C., Papazisis, K., Venizelos, V., Xanthakis, I., Xepapadakis, G., Banu, E., Eniu, D. T., Negru, S., ... Nasioulas, G. (2019). Analysis of hereditary cancer syndromes by using a panel of genes: Novel and multiple pathogenic mutations. BMC Cancer, 19(1), 535. https://doi.org/10.1186/s12885-019-5756-4

Udler, M. S. (2019). Type 2 Diabetes: Multiple Genes, Multiple Diseases. Current Diabetes Reports, 19(8). https://doi.org/10.1007/s11892-019-1169-7

Utinans, A., & Ancane, G. (2014). Belief in the paranormal and modern health worries. SHS Web of Conferences, 10, 00048. https://doi.org/10.1051/shsconf/20141000048

Veltri, G. A., & Suerdem, A. K. (2013). Worldviews and discursive construction of GMO-related risk perceptions in Turkey. Public Understanding of Science, 22(2), 137–154. https://doi.org/10.1177/0963662511423334

Chapter 11:
Brym, R. J. (2019). SOC+.

Brym, Robert. 2018. "The Social Bases of Cancer." Pp. 81–102 in Robert Brym, Sociology as a Life or Death Issue, 4th Canadian ed. Toronto: Nelson.

Condit, C. M., Ofulue, N., & Sheedy, K. M. (1998). Determinism and mass-media portrayals of genetics. American journal of human genetics, 62(4), 979–984. https://doi.org/10.1086/301784

Entine, J. 2000. Taboo: Why Black Athletes Dominate Sports and Why We Are Afraid to Talk about It. New York: Public Affairs.

Gibbons, A. (2017, December 10). Bonobos Join Chimps as Closest Human Relatives. Science Mag. https://www.sciencemag.org/news/2012/06/bonobos-join-chimps-closest-human-relatives#:~:text=Ever%20since%20researchers%20sequenced%20the,them%20our%20closest%20living%20relatives.

Haran, J., & O'Riordan, K. (2018). Public knowledge-making and the media: Genes, genetics, cloning and Mass Observation. European Journal of Cultural Studies, 21(6), 687-706.

Heine, S. J., Dar-Nimrod, I., Cheung, B. Y., & Proulx, T. (2017). Essentially biased: Why people are fatalistic about genes. In Advances in experimental social psychology (Vol. 55, pp. 137-192). Academic Press.

Hughes, E., & Kitzinger, J. (2008). Science fiction fears? An analysis of how people use fiction in discussing risk and emerging science and technology.
Michael, M., & Carter, S. (2001). The Facts About Fictions And Vice Versa: Public Understanding of Human Genetics. Science as Culture, 10(1), 5–32. https://doi.org/10.1080/09505430020025483

Molla, R. (2019, December 13). Genetic testing is an inexact science with real consequences. Vox. https://www.vox.com/recode/2019/12/13/20978024/genetic-testing-dna-consequences-23andme-ancestry.

Resnik, D.B., Vorhaus, D.B. Genetic modification and genetic determinism. Philos Ethics Humanit Med 1, 9 (2006). https://doi.org/10.1186/1747-5341-1-9
Sawyer, Diane. 2015. "Bruce Jenner—The Interview." 20/20 24 April. Retrieved May 3, 2021 http://abc.go.com/shows/2020/listing/2015-04/24-brucejenner-the-interview).

Seacrest, R., Murray, J., Goldschein, G., & Jenkins, J. (2021, April 1). Winners Take All. Keeping up with the Kardashians. episode, Los Angeles, California; E!

Simpson, J. A., and Michael Proffitt. "Popular Culture." Oxford English Dictionary, Clarendon Press, 1997.

Spielberg, S. (1993). Jurassic Park. Universal Pictures.

Turney, J. (2000). Frankenstein's footsteps: science, genetics and popular culture. Yale University Press.

Vackimes, S. Mutant, Hero or Monster? Genetics in Cinema. Part I. A Broader Conception of Heredity, 137.

Young, Rob. 2016. "Canadian Media Usage Trends Study 2016." Retrieved 13 January 2018 (https://iabcanada.com/research/cmust).

Chapter 12:

Arnaud, C.H. (2017). Thirty years of DNA forensics: How DNA has revolutionized criminal investigations. Chemical and Engineering News. https://cen.acs.org/analytical-chemistry/Thirty-years-DNA-forensics-DNA/95/i37.

Bu, Q. (2019). Reassess the Law and Ethics of Heritable Genome Editing Interventions. Issues in Law & Medicine, 34(2).

Kedmey, D. (2015). The promising and perilous science of gene editing. TED. https://ideas.ted.com/the-promising-and-perilous-science-of-gene-editing/.

Ledford, H. (2019). CRISPR babies: when will the world be ready? Nature. https://www.nature.com/articles/d41586-019-01906-z#ref-CR9.

Mandip, K.C., Steer, C.J. (2019). A new era of gene editing for the treatment of human diseases, Swiss Med Wkly, 149. https://doi.org/10.4414/smw.2019.20021.

National Academy of Science, Engineering and Medicine. (2015). On Human Gene Editing: International Summit Statement. https://www.nationalacademies.org/news/2015/12/on-human-gene-editing-international-summit-statement.

OntarioGenomics. (2014). Personalized Medicine: The Future is Now [video]. YouTube. youtube.com/watch?v=-HruX2tDf7E&t=72s.

Pang, R. T. K., Ho, P.C. (2016). Designer babies. Obstetrics, Gynecology and Reproductive Medicine, 26 (2).

Portin, P. (2014). The birth and development of the DNA theory of inheritance: sixty years since the discovery of the structure of DNA. J Genet, 93, 293–302. https://doi.org/10.1007/s12041-014-0337-4.

Pranke, I., et al. (2019). Emerging Therapeutic Approaches for Cystic Fibrosis. From Gene Editing to Personalized Medicine, Front. Pharmacol, https://doi.org/10.3389/fphar.2019.00121

Thulin, L. (2019). Four U.S. CRISPR Trials Editing Human DNA to Research New Treatments. Smithsonian Magazine. https://www.smithsonianmag.com/science-nature/four-us-crispr-trials-editing-human-dna-for-new-medical-treatments-180973029/.

Zhou, Qi. (2020). Human embryo gene editing: God's scalpel or Pandora's box?, Briefings in Functional Genomics, 19(3), 154–16. https://doi-org.myaccess.library.utoronto.ca/10.1093/bfgp/elz025.

N.A. (2016). Discoveries in DNA: What's New Since You Went to High School? Saint Louis University. https://www.slu.edu/news/2016/august/Eissenberg-genetics-essay.php.

N.A. (2017). Preimplantation Genetic Diagnosis: PGD. American Pregnancy Association. https://americanpregnancy.org/getting-pregnant/infertility/preimplantation-genetic-diagnosis-70971/#:~:text=Preimplantation%20genetic%20diagnosis%20(PGD)%20is,passed%20on%20to%20the%20child

www.ingramcontent.com/pod-product-compliance
Lightning Source LLC
Chambersburg PA
CBHW030119170426
43198CB00009B/676